International Conference on Design & Construction of Wind Farm Support Vessels 2017

London, United Kingdom
29-30 March 2017

ISBN: 978-1-5108-8300-0

Printed from e-media with permission by:

Curran Associates, Inc.
57 Morehouse Lane
Red Hook, NY 12571

Some format issues inherent in the e-media version may also appear in this print version.

Printed with permission by Curran Associates, Inc. (2020)

For permission requests, please contact The Royal Institution of Naval Architects at the address below.

The Royal Institution of Naval Architects
8-9 Northumberland Street
London, WC2N 5DA
United Kingdom

Phone: 020 7235 4622
Fax: 020 7259 5912

publications@rina.org.uk

Additional copies of this publication are available from:

Curran Associates, Inc.
57 Morehouse Lane
Red Hook, NY 12571 USA
Phone: 845-758-0400
Fax: 845-758-2633
Email: curran@proceedings.com
Web: www.proceedings.com

CONTENTS

The Royal Institution of
Naval Architects

INTERNATIONAL CONFERENCE

Design & Construction of Wind Farm Support Vessels

29-30 March 2017, London, UK

HYBRID AND ELECTRIC POWER FOR WIND FARM SUPPORT VESSELS – BRINGING CLEAN ENERGY FULL CIRCLE

G Brown, PBES, Canada

SUMMARY

Energy storage technology is a major part of the current evolution of the maritime industry; hybrid and fully electric systems have successfully been installed on a variety of commercial vessels including ferries, offshore support vessels, fish boats, and tugboats. While environmental regulations have helped to create demand for this sustainable technology, the primary market driver is increased safety and reliability of the vessels and a significant decrease in operational and fuel costs.

As a result, there is now a new set of decisions and system options for vessel owners to consider that effect system lifespan, ROI and performance. This paper will examine the differences between hybrid and fully electric systems, look at considerations for new versus retrofit system design, as well as new energy storage innovations such as PBES' CellSwapTM refurbishment system.

1. INTRODUCTION

The European offshore wind farm industry is booming. In 2015, 3,10MW of grid-connected capacity was added[1], 108% more than in the last 10 years! With this growth comes a need for evolution in service vessels that support the industry.

A great challenge for the support vessel sector is to understand what the industry will look like in coming years. With an increasing progression in design and technology, as well as demand from operators, ship owners need to make decisions today that are useful for the next 10-15 years.

Energy storage technology is a major part of the current evolution of the maritime industry; hybrid and fully electric systems have successfully been installed on a variety of commercial vessels including ferries, offshore support vessels, fish boats, and tugboats. While environmental regulations have helped to create demand for this sustainable technology, the primary market driver is increased safety and reliability of the vessels and a significant decrease in operational and fuel costs.

As a result, there is now a new set of decisions and system options for vessel owners to consider that effect system lifespan, ROI and performance. This paper addresses new considerations for low and zero emission vessels such as:

Hybrid vs. Full electric options:

- Identifying operational profiles such as range and expected use
- Battery size/cost
- Refit an existing vessel or build new

This paper will examine the differences between hybrid and fully electric systems, look at considerations for new versus retrofit system design, as well as new energy storage innovations such as PBES' CellSwap^{TM} refurbishment system.

2. WIND FARM SUPPORT VESSEL (WFSV)

Rough weather and high waves often characterize the working environment for people and equipment during the installation, and maintenance of offshore wind farms. This means that wind farm services vessels have extraordinary requirements for immediate power, endurance, robustness and safety in order to maintain their operational duties in all weather. A service vessel requires powerful bollard pull capabilities, and in general excellent sea keeping abilities to withstand wind, waves and currents. In these demanding environments, human safety as well as operational expenditures is of key concern, thus service vessels must be built to the highest standards.

To make sure essential control and machinery systems are reliable during ship operations, redundant systems are key. By using energy storage as redundancy, WFSVs will become much safer vessels, reducing risk by ensuring backup and work continuance. For smaller vessels where additional machinery will not fit, energy storage adds increased safety to the operation where they might not be able with traditional propulsion systems.

Modern lithium-ion energy storage is also capable of providing high power for extreme conditions and demanding situations. The energy storage solution can respond much faster than diesel mechanical solutions to changing weather or conditions. The response time is in milliseconds and allows the system to shave peaks as they happen, decreasing wear and tear of the engines and reducing fuel consumption. Today's energy storage solutions are particularly suited to meet the demanding operating conditions and power requirements of hard working service vessels.

2.1 HOW ENERGY STORAGE SOLUTIONS IMPROVE OPERATIONS

Wind farm vessels spend the majority of their time stationary at the wind farm, and generally do not exceed 6-8 knots while amongst the turbines. This means energy storage can often be used as the sole power supply, thereby reducing fuel and maintenance costs as well as improving working conditions. As the crews are working at the turbines, there will be no noise, air pollution or vibration from the vessel.

3. FULL ELECTRIC VESSELS

Full electric vessels have already been proven to work in heavy industrial applications, and with rapidly increasing technology advancements, full electric operation is more feasible today than ever. With increased capacity, as well as state of the art safety mechanisms, vessels can potentially operate on grid power alone. This means further reduced operational costs and zero environmental impact. Ferries in Northern Europe have been operating on battery power from the grid for years, proving the ability to handle significant power and energy demands. With current energy storage solutions able to provide higher discharge rates, the total installation can be smaller and lighter, which is particularly relevant for vessels that are run on higher speeds.

When all the energy is coming from batteries alone, demands on the system are very high, and cooling systems are essential to maintain performance and increase battery life. Liquid cooling will allow discharge rates as high as 3C, without damage to the battery.

4. FULL ELECTRIC WIND FARM VESSELS

Full electric vessels usually require fixed power and cycle demand, and regular access to charging facilities (similar to ferry operations). Wind farm vessels often have unpredictable operational profiles, run on high speeds, and have no access to charging offshore, therefore full electrification is unlikely today.

5. HYBRID WIND FARM VESSELS

Hybrid systems offer more flexibility for dynamic operational profiles. When used on a wind farm support vessel, the energy storage solutions is charged using shore power overnight. The vessel will transit through the harbour on battery power, operating in zero emissions mode. When high-speed transit is needed to reach the wind farms, the diesel engine/generator takes over. Many vessels have hours of waiting and loitering in the wind farms on standby. This low speed loitering, in a typical system, is done using inefficient load cycle for diesel engines. Inefficient loading leads to increased maintenance and reduced engine life. In a hybrid, this part of the operational profile is powered by the energy storage solutions. An added benefit is a quiet, vibration and emission free environment for vessel workers.

Batteries provide a "bridge", optimizing generators and reducing low load operation. This offers instant spinning reserve and clean, quiet low speed propulsion and house power. The result is decreased fuel consumption, air pollution, noise and maintenance costs.

6. OPERATIONAL PROFILE

Below is a typical duty cycle and standard vessel characteristics for an example wind farm support vessel that might be found working offshore.

Vessel: 20 m catamaran
Tonnage: 40
PAX: 12
Engine: 2x600 kW
Battery size: 390 kWh
Travel speed: 25 knots
Idling at wind farm: 6-8 knots

In this example, the vessel would typically spend 2 hours sailing to the field, 10 hours working, and 2 hours back again. The vessel goes out to the wind farm at 25 knots. When in operation at the wind farm, it would run solely on battery power at 6-8 knots or at idle, saving roughly 10 hours of idling run time per engine per day of operation, see figure 1.

Operation	Time	Energy
Trip to wind farm	2 hours	Fuel engine
At wind farm	10 hours	Battery electric
Return to shore	2 hours	Fuel engine

Figure 1: Typical duty cycle of wind farm support vessel.

6.1 TWO POSSIBLE SOLUTIONS

Below is a typical duty cycle and standard vessel characteristics for an example wind farm support vessel that might be found working offshore.

6.1 (a) Parallel Hybrid

A parallel hybrid is when both a diesel engine and electric motor are being used alternately in the drive-line. In the parallel hybrid WFSV, the vessel runs on a diesel engine with mechanical drive during the trip to from, the wind farm. At the wind farm or during low speed operation, the diesel engines are shut down and the vessel runs on batteries powering electric drives. This low speed operation is without noise, vibration or pollution. Further, turning off the engines for long periods at the farm significantly reduces wear and tear of the engines and associated maintenance costs.

6.1 (b) Serial Hybrid

A serial hybrid uses an electric motor solely for propulsion, with energy directly from the energy storage solutions or from a diesel generator as required. In the serial hybrid example a vessel might have two generators of 600kW each. The batteries shave peak loads in the operation to ensure the diesel engines are running at optimal loading for the entire duration of the duty cycle. low speed operation is without noise, vibration or pollution. Further, turning off the engines for long periods at the farm significantly reduces wear and tear of the engines and associated maintenance costs.

7. BENEFITS OF HYBRIDIZATION

The cost of energy storage in wind farm vessels must be determined on a project basis, but in the example above, the reduced fuel consumption and maintenance costs will achieve return on investment very quickly. Further, the environmental savings are excellent in that the vessel will spend far less time in idle/low load conditions where diesels are at their least efficient. The crew will benefit also; the electric motors run with less noise, vibration and have no fumes. Anecdotal reports show far less crew fatigue at end of shift, which may result in fewer lost man hours, fewer injuries and less risk to the machinery.

8. REFIT AN EXISTING VESSEL OR BUILD NEW

Energy storage can work well both in refit and new build projects. Considering the number of recently built WFSV's currently in operation, there is great opportunity to retrofit existing vessels. It makes sense to refit a newer hull with energy storage at standard maintenance intervals when the engines are being rebuilt or overhauled.

The savings from energy storage are greater the longer the vessel runs as a hybrid as the reduced operational costs are in effect for a longer portion of the vessel life. In a similar mind-set, the costs are lower to build a hybrid vessel when it is designed that way from the beginning and it spends its entire lifespan as a hybrid, thus achieving payback much quicker.

9. SAFETY

Marine propulsion batteries are commonly hundreds of times as large as automotive electric vehicle batteries. The large size, high energy density, rapid charging requirements and extreme operational patterns represent new challenges in relation to safety. In order to prevent fire or thermal runaway, several measures are implemented. The first line of defence is thermal management. Controlling temperature with a liquid cooling system in the core of the battery is integral to ensuring not only the safety of the system but also the longevity. The cells contained in the battery last far longer when kept at an optimal temperature and a robust cooling system can actually prevent thermal runaway rather than reacting to it after it has already occurred. Cooling, fire suppression and battery management systems are key to providing safety and protection from the cell level up to the system level.

While the industry standard for many years was passive air cooling, it is increasingly apparent that the smaller systems demanded by industry are required to operate at, or beyond, the limits of passive cooling. Virtually all of today's energy storage solutions employ some form of liquid cooling, either as an optional addition to the standard system or as an integral component. Advanced, state of the art batteries use integral cell level cooling systems; the coolant circulates within the very core of the battery module at low pressure thus enabling greater charge and discharge currents, increased lifespans and reduced system sizes. Typically, the energy storage solution will connect to a standard chiller using an inexpensive circulation pump and is able to meet the very high power and load demands with a far smaller system size and resulting cost savings.

Given the possible issues associated with fire and explosion, the class groups and flag authorities rightfully been very focussed on how to prevent and manage fires and thermal runaway. No matter the amount of regulation applied to suppression, it does not remove the battery manufacturers from the responsibility of incorporating sophisticated prevention systems into the design of the batteries. With lithium energy storage solutions now regularly being discussed that exceed several MWh of capacity, the risk of thermal runaway or fire cannot be taken lightly. Today's hybrid designs must take this into account and do everything possible to ensure that a fire cannot start in the first place. As previously mentioned, this has created a shift in thinking that is driving designs to incorporate high performance liquid cooling systems. These liquid cooling systems manage battery safety inside the core of the module through temperature control and management at the cell level. Fire suppression is critically important but must be viewed as a secondary system to manage the issue in extreme circumstances, after all else fails.

Fire suppression systems therefore are recommended to control external fires adjacent to the energy storage solution to prevent them from causing a thermal event in the battery room. If desired, fire suppression in the battery room may also be employed to further give peace of mind as a backup system. Fog type fire suppression systems provide adequate cooling to suppress virtually any fire (outside of a major catastrophe involving the ship itself) that may pose a hazard to the energy storage system. In order to meet class standards, the energy storage system itself must be rated for IP67 water resistance and therefore able to safely withstand activation and use of fire suppression.

10. BATTERY MANAGEMENT SYSTEM

The Battery Management System (BMS) is able to actively monitor the state of health of the system within the temperature warnings, if a specific component in any one part of the entire system is out of spec, the system will warn the captain and the team who is monitoring it. The monitoring team will then proactively engage with the vessel and determine what, if any course of action need be taken. If the warnings continue without intervention from the team, or if the vessel crew ignores the warnings, the system will protect itself and the vessel by disengaging from the DC bus and isolating all the modules in the system via their internal contactors, effectively reducing system voltage from a maximum of 1000V to ~100V (the voltage of a single module). As the controls are powered separately from the energy storage solution, they are safer in that there is redundancy in the system. It will always have an external power source ensuring the cooling system is operating and the management system is communicating with the vessel and system administrator team at all times, regardless of the system status.

11. CHARGING INFRASTRUCTURE

The hybrid wind farm support vessel could be connected to shore power over night to fully charge. The vessel can leave the dock and harbour on battery/electric power, which also means zero emissions. Later in the day diesel power is used for the high-speed transit, during this time the batteries are re-charging from on-board generators. In some cases, there may be opportunity to recharge on-board batteries from other larger vessels that have a surplus of power available. This form of opportunity charging may in fact make the idea of a fully electric WFSV a reality in very short order.

The modern hybrid or electric energy storage solution requires the ability to communicate with and control the charging infrastructure of the batteries. A lithium battery is very sensitive to voltage and current. Voltage must be kept at a constant setting specified by the manufacturer. The BMS must be able to direct the charging system to increase current at specific set points to increase charging rates, ultimately to decrease charge times and optimize usefulness of the system. Conversely, the BMS must also be able to reduce current at other times to increase longevity of the system and meet the vessel owner's objectives for lifespan of the system. The BMS will take all types of information from the modules and surrounding systems, such as temperature, state of charge and system age, to determine the best charging profile at any given time.

12. THE FIVE-YEAR VS. TEN-YEAR SYSTEM

Vessel energy storage solutions can now be designed based on a 5-year lifespan, compared with the traditional 10-year lifespan approach, which is typical of the industry. Even though energy storage is proven to reduce fuel costs, and have a short payback time, the initial investment can be substantial. With a smaller system, the investment is inherently reduced and ship owners can enjoy an even faster payback time. Instead of investing for 10 years, ship owners can invest in the next five, and at the end of those five years the cells within the energy storage solution are replaced and the modules are "re-cored" and upgraded. A key condition for the 5-year battery is to have a highly effective cooling system to control temperature. With liquid cooling of each cell, it is possible to achieve maximum charge and discharge of the battery without compromising safety or accidentally depleting the battery

The 5-year battery system is a new approach to marine energy system sizing, it significantly reduces operational and installation cost, and avoids the loss of system life and capacity due to lithium cell chemistry aging, which on a 10-year system, can account to more than a 10% loss. As soon as cells are manufactured they start to age based on chemical reaction. Depending on the cell chemistry and how it's stored, it may lose 1-2% per year. Similar to compound interest, the longer the system life, the more a system will lose over time. On a 10-year system, that 2% can amount to over 20% capacity. With a 5-year system cell capacity loss is reduced greatly.

The physical footprint of any energy solutions is an important consideration for vessel designers and operators. When choosing the appropriate system type, a significant advantage of the 5-year battery is the greatly reduced physical size and weight. As of 2017, we have seen system size reductions in the order of 30% .The reduced size will also save weight allowing for higher speed at less fuel consumption. For many ship owners, especially for fast going vessels like wind farm support vessels and fast ferries, vessel weight is of key concern. Fast going vessels are often lighter , which can put restrictions on energy storage installations, which are relatively heavy. With a smaller energy storage solution, weight is reduced, and operators can benefit from a high-power system that improves working conditions, operations and costs.

13. CELLSWAP™ PROGRAM

Modern battery systems for hybrid and full electric vessels are becoming even more efficient and cost effective with recent advent of module re-coring. With a 5-year battery design, a much smaller battery footprint and weight will result. At the end of life, cells within the energy storage system are replaced and the modules are "re-cored" and upgraded. The re-coring program is a unique feature of advanced energy storage solutions that will revolutionize the way ship-owners use this technology. The concept is based on only replacing the cells of the battery system, reducing the total lifetime cost of the system. Essentially this means that instead of changing the entire energy storage solution, it is possible to only change the depleted cells.

Similar to rebuilding a diesel generator, a re-coring program only replaces specific elements of the battery system, while the majority of the installation remains intact. The ability to swap out just the cells means that almost no components are thrown away or need to be recycled, greatly reducing waste.

CellSwap™ is possible as the system infrastructure is made from high quality materials and has a longer lifetime than the cells. By this, waste is reduced and money is saved. Another key attribute to a CellSwap™ program is that there are no changes in shipboard integration and power management. A new battery system would require re-integration where costs are difficult to assess and control. The ability to re-use existing systems also ensures there are less issues with legacy/obsolete components and software.

With the rapid progress in energy storage solution development, many ship-owners do not want to risk having invested in yesterday's technology. With CellSwap™, the operator will be able to take advantage of the latest cell technology, which will enhance performance, reliability and safety of the battery system. With advances in cell technology and performance customers can always be on the leading edge of performance for the life of the vessel.

13.1 PBES CELLSWAP™ PROCESS

The process is simple and safe, as the cell stack easily separates from the electrical controls. The new stack of cells is pre-made in the factory and shipped to the vessel for installation by certified technicians. The front end of each module containing the electronics is detached and the core which houses the cells is replaced with the new stack.

PBES CellSwap™ plan is a replacement of all consumable elements at the 5-year lifespan of the original system, including cells and any other parts to be replaced at that time. Regular scheduled maintenance is performed on the control systems at the same time. A PBES Service Technician then performs an on-site FAT, a re-commissioning and CAT on the re-cored battery system. Before committing to a plan, the existing battery system and working environment is to undergo a full inspection by PBES Service technician. Consultation with the vessel as per use plan for the year that the batteries will be replaced will also be required.

With environmental concerns an important part of many business' plans that must reflect the need for sustainability and accountability, the waste reduction from CellSwap™ is a key benefit.

13.2 ENVIRONMENTAL FOOT PRINT

Traditional battery systems require the entire system to be replaced at end of life. This leads to a great deal of industrial and electrical waste. While many of the components of the battery may be recycled, the cost and pressure on existing recycling infrastructure is significant. The volume will only increase as the new fleet of hybrid and full electric vessels begins to age and more battery systems are replaced.

14. LIMITATIONS AND CONSIDERATIONS

The re-coring process does not work for all batteries. The smaller 5-year system can only be employed through the use of an energy storage solution that is capable of being run at far higher discharge rates. This results in increased stress due to additional heat and cycles, which in turn ages the batteries faster and places greater demands on safety systems. The well documented tendency for lithium batteries to enter thermal runaway when abused, means robust cooling and thermal runaway suppression systems must be integrated into the system.

15. REFERENCES

References should be numbered in the text where they occur, with the number in square brackets [1] and a complete list given at the end of the paper.

Cross Referencing, footnotes, endnotes and 'Smart Tags' must NOT be used for figures and/or references.

1. Ho, Andrew., 'The European offshore wind industry - key trends and statistics 2015', *A report by the European Wind Energy Association*, 2016.

16. AUTHORS BIOGRAPHY

Grant Brown holds the current position of Vice President Brand and Marketing at Plan B Energy Storage. His 8 years of experience working in the marine and industrial energy storage sector provide him with unique insights to market trends and developments. His previous experience includes helping found the first company to develop a purpose designed energy storage system for commercial marine hybrid and electric propulsion.

THE HOUR OF POWER - HYBRID MARINE TECHNOLOGY FOR WFSV APPLICATIONS

J Haynes, Shock Mitigation Ltd, UK

SUMMARY

If wind, wave and tidal energy installations are striving for genuine 'green' credentials it is logical to reduce consumption of fossil fuels wherever possible. Maritime sectors with predictable duty cycles are potentially well suited to hybrid diesel / electric systems. Key drivers for hybridisation of Wind Farm Support Vessels include reduced emissions, green port compliance, reduced fuel costs, reduced engine maintenance costs, redundancy, reduced noise and vibration at low speeds.

The Hour Of Power concept enables vessels to run in and out of port for an hour on electric with battery power, then carry out their open sea work on diesel power. The Hour Of Power concept lends itself to WFSVs operating in the ongoing wind farm maintenance phase. The aim of this hybrid solution is to enhance conventional power and propulsion systems. Vessels can reduce emissions and improve fuel consumption, whilst extending engine maintenance periods and overall engine life.

NOMENCLATURE

WFSV: Wind Farm Service Vessel
IMO: International Maritime Organisation
EEDI: Energy Efficient Design Index
LLL: Load Line Length
CTV: Crew Transfer Vessel
COTS: Commercial Off The Shelf
PHEV: Plug-In Hybrid Electric Vehicle
E&P: Exploration and Production
MCA: Maritime Coastguard Agency
DNV-GL: Det Norske Veritas - Germanischer Lloyd

1. INTRODUCTION

In 2015 two significant developments caused many owners, operators and builders of professional vessels to consider hybrid marine power and propulsion. Firstly the new emissions laws in ports and secondly there is now an incentive for high technology manufacturers to invest in developing highly efficient batteries and alternative marine energy systems. Innovative companies such as Tesla rely on increased battery performance and reduced battery costs to accelerate mass market adoption of electric automobiles and domestic battery power. As these efficiencies move from concept to production other transport sectors are collaborating to share technology as they evolve their own next generation power options.

2016 saw a reduction in oil prices due to global politics linked to supply and demand. Although low oil prices extend the return on investment / cost recovery period for hybrid systems the interest in hybrid has continued to grow. Diesel engines are designed to run at around 70% to 80% of maximum power. Hybrid systems that introduce batteries, or utilise smaller generators, increase efficiency when vessels are loitering or moving slowly at low engine loads that are inefficient for diesel engines. Analysis of engine data on WFSV can be utilised to identify potentially inefficient duty cycles. The growing evidence linking diesel emissions to human health is a driver to reduce air and water pollution around ports.

2. OFFSHORE WIND ENERGY

Offshore wind farm development has been led globally by the UK, followed by Germany. Both countries have governments with long term commitments to renewable energy. Denmark, Belgium and The Netherlands are building wind farms, but they are a long way behind the leaders in offshore wind. In 2014 the UK accounted for over 50% of all the offshore wind energy generated globally. All of these countries border the North Sea, a notoriously stormy sea area, which has a high frequency of wind to drive the turbines.

The coastal countries of northern Europe are all experienced maritime nations with access to innovative design, high technology and construction expertise that has helped to develop the vessels that support offshore wind. The latest vessels have been optimised for high speed passage, turbine access and loitering in harsher conditions. These designs are intended to have the capability to be utilised in sea areas such as the German North Sea and UK Round 3 wind farm developments that are further offshore. The last ten years has seen the wind farm service vessel market expand significantly with a steady evolution in design as operators aim to become more competitive and meet changing requirements. Key drivers for new designs are the ability to access wind farms in higher sea states and improved fuel efficiency.

2.1 OFFSHORE WIND FARM OPERATIONS

The first stage of wind farm installation typically involves heavy lift construction vessels and construction support vessels. Tasks for installation vessels include the transfer of support structures and turbines offshore. On site these vessels establish bases for lifting and installation operations. These vessels can also provide offshore access and accommodation for crew. This phase of the installation process includes transportation of the nacelle, hub assembly, blades, towers and array cables from construction port to site.

Once the wind farm construction phase is underway vessels on site include Wind Farm Service Vessels (WFSV) which can overlap roles with Crew Transfer Vessels (CTV). The multi role WFSV enables fast personnel transfer with the capability for utility work such as moving equipment, delivery of spares, enforcing safety zones, conducting environmental studies and providing support for divers.

Once a wind farm is completed and connected to the grid preventative and corrective maintenance are ongoing tasks for WFSVs. The schedule of visits to turbines includes routine surveys and inspections. The aim is to keep the entire wind farm ready to generate wind for the maximum number of days per year. Preventative maintenance programmes are designed to decrease the number of failures, limit downtime and extend the life of the turbine, foundation and associated components.

2.2 WFSV CHARACTERISTICS

Current fleets of WFSVs have been specifically designed to work in the wind farm support sector. As catamarans offer reduced resistance to motion and reasonable stability they are now the hull form of choice for most WFSV working in northern European waters. As the demand for wind farm support has intensified, vessels have been specifically developed for transiting to wind farms in a wide range of conditions then delivering technicians onto wind farm turbines.

In the early stages of wind farm development vessels ranged from 15 to 18 metres (50 to 60 feet). Due to the International Load Line convention vessels are typically delivered below 24 metres LLL (Load Line Length). In recent years the majority of vessels have been designed to 24 metre (80 feet) with wide beams to get maximum working space and payload onboard.

Figure 1: 24m WFSV operated by Seacat Services. Credit: Alicat / South Boats IOW

A 24 metre catamaran may be required to carry up to 30 tonnes of cargo. When moving technicians transit speeds range between 20 and 25 knots with some vessels having a top speed of 30 knots. Fixed pitched propellers are used to drive around 50% of WFSV. Water jets are used on around 40% of the current fleet. Water jets offer good manoeuvrability and the shallow draft can be essential on some offshore wind farm sites. Other propulsion systems include controlled pitched propellers (CPP) with various new systems under trial and evaluation.

These specialist designs comprise of a pair of hulls reinforced by substantial vessel superstructure and high bridge position, which gives good line of sight for the captain when embarking personnel or cargo from the bow. Vessels have evolved with improved underwater hull shapes to reduce engine power requirements and reduce fuel costs, which are a significant part of operation and maintenance costs. When boats are chartered to a wind farm the boat owner provides vessel and crew for a fixed period with fuel costs usually borne by the wind farm operator.

2.3 TRANSITS & TURBINE TRANSFERS

Transit time and accessibility to the wind farm are often weather dependent. In the North Sea, sites further from shore typically experience larger significant wave heights. Transferring maintenance personnel and their equipment from vessel to wind turbine safely in various sea states is a major challenge in offshore wind farm operations. In these locations the turbine foundation is fixed to the seabed resulting in considerable relative movement between the vessel and the structure.

Figure 2: 26m WFSV operated by Seacat Services. Credit: Alicat / South Boats IOW

Turbine transfers are where a vessel uses engine power to push the bow against 'j-tubes' which run vertically on the outside of the turbine access ladder. Thrust is applied to increase friction and reduce motion between specially designed bow fenders and the landing platform. Maintenance technicians step onto the turbine access ladder which is set back from the j-tubes providing a safety zone to prevent personnel on the ladder from being crushed. Large waves, especially when coupled with strong currents, can cause the vessel to lose position or contact with the turbine.

3. CLIMATE CHANGE & SHIPPING

In December 2015 the Paris COP21 (Conference Of the Parties) conference on climate change resulted in 196 nations agreeing to reduce their overall greenhouse emissions and endeavour to limit global warming to less than 2 degrees Celsius compared to pre-industrial levels.

As our planet progressively transitions from fossil fuels to alternative energy sources there is a need for self sufficient land and ocean based grid applications. The consistent challenge for genuine 'green energy' is extending the capabilities of energy generation and storage technologies.

3.1 LAND TRANSPORT IS CHANGING

Reducing emissions from cars, busses and trucks in the world's major cities has been a major driver for efficient electric propulsion and battery power storage. Automotive manufacturers in Europe, the Far East and the US have recognised that hybrid technologies such as PHEV (Plug-in Hybrid Electric Vehicle) utilising serial production battery technologies, including lithium ion, will be dominant for the next decade. Conventional power technologies are trying to defend their position and the manipulation of automotive diesel emissions data that surfaced in 2015 has increased the global awareness of emissions and accelerated the commercial technology drivers for alternative energy in all transport sectors.

3.2 ECA EMISSION CONTROL AREAS

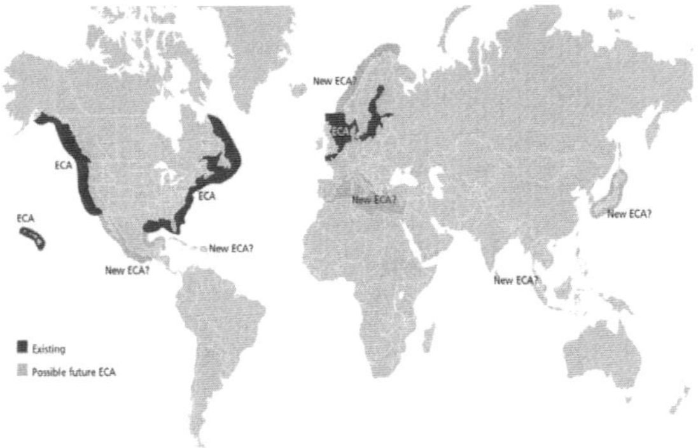

Figure 3: Emissions Control Areas - Global

1 January 2015 - ECA Zone requirements (IMO Annex VI) came into effect for USA, Canada, North Sea, Baltic. The maritime sector can be slow to adopt changes, but the ocean going shipping community could not ignore the 2015 emissions regulations. Options to reduce emissions include exhaust cleaning 'Scrubbers'.

Some operators that infrequently visit ECAs may make a commercial decision that it is cheaper to be fined than to comply. However this is not a viable option for ferries and cruise ships that make regular visits to ports.

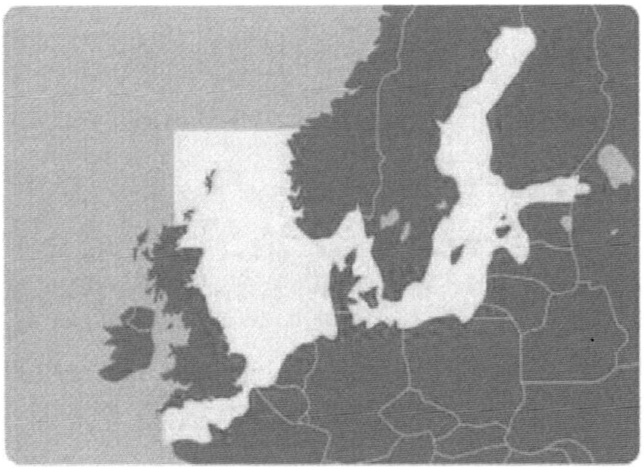

Figure 4: Emissions Control Areas - North Sea & Baltic

The implications are that certain types of vessels may be banned from entering ports around the world simply, because of their diesel exhaust emissions. This will increase the cost of world trade as a high percentage of all goods are transported by sea.

3.3 ENVIRONMENTAL INCENTIVES

The adoption of hybrid technology for vessels operating in ports with large conurbations of people around them, may be driven by national / regional environmental penalties or forward thinking environmental incentives.

In Norway government recommendations specify zero or low-emission technology in all domestic public ferry transportation. The overall aim is to lower emissions of greenhouse gasses and to become a leading supplier of zero emission solutions.

Norway has taken the following actions to increase the volume of zero emission cargo transport by sea:

- Investment support for shore power in ports.
- Increased grants from NOx fund for battery installations.
- Strong reduction of duty on electric power to ships in ports.
- Grants for ship investments with reduced power consumption.
- Zero emission solutions required in government ferry tenders.
- Research and development grants targeting the maritime industry.

Source: European Shortsea Network

In Sweden the Port of Gothenburg rewards vessels that are environmental high performers. The environmental discount on port charges is based on two separate environmental indexes: the Environmental Ship Index

(ESI) and Clean Shipping Index (CSI). In 2016 vessels that had a score of at least 30 according to ESI, or which are classified as green according to CSI, received a 10% discount on port charges.
Source: Port of Gothenburg

3.4 FUEL FOR SHIPPING IS CHANGING

Dramatically reducing pollution in both water and air, particularly in ports and around people, could be the most significant driver for change in the maritime sector since coal and steam gave way to fuel oil. Emissions from fossil fuels come at a price and professional mariners know the legislation relating to marine pollution (MARPOL) plus the serious financial implications of getting it wrong.

The diesel in ships tanks is now getting so light and highly treated that navies are able to run their RHIBs and boats on diesel from the mother vessel. While conventional oil based fuels remain the main fuel option for most in service vessels, Liquid Natural Gas (LNG) is now a proven and available fuel solution for ship propulsion. The commercial opportunities of using LNG are now gaining interest for new builds and conversion projects, however in many parts of the world there is limited infrastructure for obtaining fuel. If LNG does not fit their needs ship operators may consider installing hybrid power systems alongside of their diesel engines.

3.5 DIESEL / ELECTRIC SYSTEMS

Submarines have utilised stored energy for many years. Diesel / electric systems have been used in large ships for many years but these are not hybrid systems. The diesel / electric vessel uses its engines to connect directly to an electrical generator, the power in the system is then transferred electrically to the propeller shaft via a motor controller and electric motor. The system may have multiple generators and multiple motors. By strict definition this is not a hybrid as there is no storage of electric energy.

3.6 SPEED & THE BALANCE OF POWER

The traditional operating model for shipping is moving goods from point to point as fast as possible. When 'slow steaming' is introduced the vessels speed is slowed down in order to lower costs by reducing fuel consumption. The requirement for 'power' is changing for many professional fast craft operations. Organisations are less focussed on overall speed and are now considering other factors during transits including the long term health of crews and the Man Machine Interface (MMI). Onboard systems cannot be operated at high speed and it is not acceptable to injure personnel during sea transits. In the offshore sector technicians need to be fit to climb a wind farm tower to do manual work at the end of a transit.

There are now more important considerations:

- Power trains that do not fail plus redundancy
- Systems that do not fail plus redundancy
- Reduced operating costs
- Extended operating range
- Improved fuel economy
- Environmental compliance in port

4. SERIAL & PARALLEL HYBRID SYSTEM

A serial hybrid is where the engine in the system only powers a generator and is not mechanically connected to the propeller shaft. A parallel hybrid is where the engine is mechanically connected along with an electric 'machine' that can operate as both a propulsion motor and a generator. The reduced electric propulsion, generator and battery demands of a parallel system reduce the costs compared to a serial system. Parallel systems are more likely to win initial market acceptance because of a perceived greater reliability, as the 'trusted' diesel engine is still connected to the propeller shaft with the electric propulsion adding a redundant system.

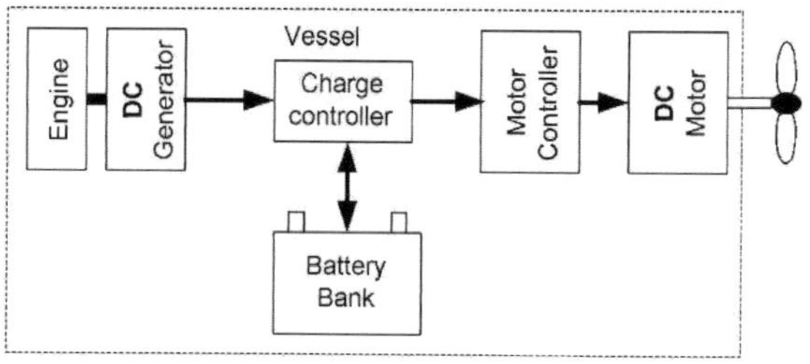

Figure 5: Serial hybrid system. Credit Hybrid Marine

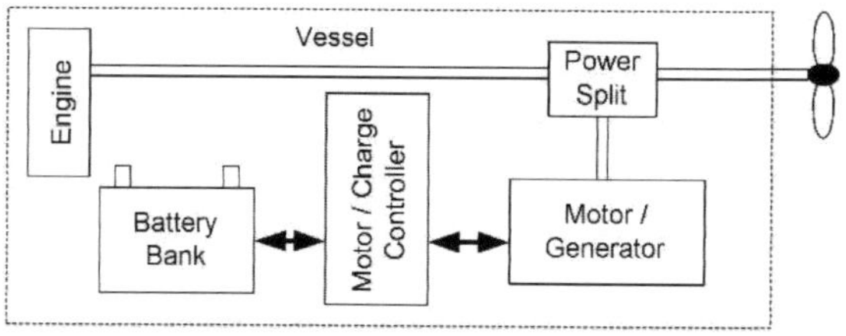

Figure 6: Parallel hybrid system. Credit Hybrid Marine

An optimised battery / electric system offers no detectable emissions, no discernible noise, a substantial reduction in fuel consumption and increased control of the vessel. For boat operators an electric motor can spin the propeller at very low speeds. Low RPM at the propeller gives improved slow speed control. Gradually increasing RPM enables operators to manoeuvre or stem a tide. This compares to repeatedly going in and out of gear with a diesel engine at idle speed, typically around 500 RPM.

5. SELECTING HYBRID FOR EFFICIENCY

When studying work cycles of vessels it is relatively straightforward to make a decision for new builds on whether to go for all electric or a diesel / electric hybrid system. For example a ferry operating over a short route with a long stopover each end could offer the perfect work cycle for 'electric only' with a land based charging system. Other issues, such as the cost of downtime and structural alterations affect viability calculations for retrofit of in-service craft.

The professional marine sector is entering a period of rapid change and commercial opportunity. Boat builders, engine manufacturers, designers and naval architects are now developing hybrid systems for WFSV, survey vessels, patrol vessels, super yacht tenders and unmanned craft. Benefits include improvements in energy reliability, increased fuel efficiency, lifecycle cost reductions and reduced emissions.

In 2015 DNV-GL USA stated that, 'energy storage is an exciting new technology, but the offshore E&P (Exploration and Production) sector of the oil and gas industry has yet to truly take advantage of it. Tugboats, workboats and OSVs are particularly suitable for hybridization.'

Not all vessels are suitable for hybridization. It is import to use a consistent process that identifies which sectors duty cycles is hybrid power most relevant for.

Quantifying and Classifying for Marine Hybrid:

- Quantify operating conditions
- Quantify the work / engine cycle
- Classify types of suitable vessels
- Classify types of 'Hybrid' systems
- Combinations of boats and 'Hybrid' systems

Typical Reasons to Consider Hybrid:

- Boats are big enough - over 10 metre (33 feet)
- Current power and propulsion is failing
- Boats run for extended time
- Boats have regular duties
- Enough budget

6. HYBRID CONCEPTS IN SERVICE

Recent projects demonstrate that hybrid systems are moving from concept and design stage to in service

6.1 HYBRID PROJECT - NORWAY

In July 2016 Norwegian shipyard Brodrane Aa delivered a 40 metre (131 feet) hybrid electric vessel to tour operator The Fjords. 'Vision of The Fjords' will carry 400 people between Flåm and Gudvangen in Norway. Cruising at up to 20 knots the ferry will use its diesel engine to reach the UNESCO world heritage area of Nærøyfjord. Once at the entrance the boat will switch to battery mode, decreasing speed to under 10 knots to reduce energy usage, with low emissions and to comply with speed limits.

The 749kw hybrid engine allows the boat to change between diesel power and emission free modes. The arrangement consists of a direct coupled diesel engine with a propeller PTO/PTI (power take off/power take in). This can disconnect the engines using the same electric generator for electric propulsion, or if needed disconnect and charge the diesel engines direct to the batteries even if no power is being put onto the propulsion system.

The vessel will feature ABB's Onboard DC Grid System that enables management and control of the energy flow between the diesel engine propeller and charging station. The Onboard DC allows the generators to run at variable speeds, making the engines more efficient and leading to fuel savings. The Grid solution will work by transferring energy to the battery during charging, which will happen at both ends of the voyage and whilst the diesel engines are running. The energy storage system, with batteries from LG Chem adapted for maritime use by Zem, has a total capacity of 576 kWh.

Hydro-electric charging stations have been built where the ferry can get its batteries fully charged in around 20 minutes. The electricity on shore is 100 per cent from hydropower and produced locally. A manual plug is used to connect the vessel to the grid.

6.2 HYBRID PROJECT - UNITED STATES

In February 2017 US shipyard All American Marine Inc. (AAM), announced the construction of a new hybrid electric 600 passenger vessel to be delivered to the Red and White Fleet of San Francisco, California. The new vessel, to be named 'Enhydra', will be the first aluminium hulled, Lithium-Ion battery-electric hybrid vessel built from the keel up under US Coast Guard SUBCHAPTER-K passenger vessel regulations and the latest guidelines for structural fire protection. The vessel will be 39 metre (128 feet) long with a beam of 9 metre (30 feet) and is specifically designed for tours of San Francisco Bay and the Golden Gate Bridge.

BAE Systems will supply their HybriDrive Propulsion System that includes a generator, control system, and AC electric traction motor. The generator will mount to a variable speed Cummins QSL9 diesel engine. The motor generator offers diesel-electric operation of the AC traction motor which is coupled directly to the propulsion shaft. With this configuration, torque is immediately available for the propeller and the speed can be precisely controlled without the need for a reduction gear.

The hybrid system will utilize battery power from two 80 kWh Lithium-ion battery packs. The Corvus Energy batteries are supplied under their Orca Energy line. The HybridDrive system can automatically utilize full electric battery operation at slower speeds and when manoeuvring in and out of the harbour. At higher speeds, the generator will automatically engage and augment the additional power demands of the traction motor.

The battery system is sufficient to meet the entire demand of the vessel's hotel load while at the same time providing silent and emission-free operation of the propulsion system.

7. THE HOUR OF POWER

Figure 7: The Hour Of Power concept utilises a combination of diesel / electric / stored energy.

Ferries with fixed schedules have the most predictable duty cycles. Various other maritime sectors are potentially well suited to 'hybrid' Diesel / Electric systems. These include pilot boats and wind farm support vessels that have relatively consistent duty cycles, often running seven days a week to drop off or collect maritime pilots and technicians. Focus is now on the sub IMO / sub 24 metre (80 feet) sectors to investigate the engineering and systems integration required to bring together viable and sustainable solutions.

The Hour Of Power is a hybrid concept, for example using a combination of Diesel / Electric / Battery, that enables a vessel to use conventional diesel engines and propulsion system, charge batteries when running diesels, charge batteries from shore power, run on Battery / Electric for up to one hour or loiter on battery for significantly longer.

The Hour Of Power focuses on hybrid solutions linked to viable business cases. This is not just green energy for the sake of it. Many commercial harbours around the world have 10 knot speed limits for low wash or safety reasons and are within 1 hour of sea. For commercial and professional organisations the simple concept of running vessels with zero emissions at up to 10 knots for one hour will shape decisions that lead to improvements of in-service systems and procurement of next generation vessels. The overall objective is fuel saving and improved efficiency by all means.

7.1 WFSV AND THE HOUR OF POWER

With vessel life cycles of over 20 years, naval architects and builders of new craft will offer designs that have space and access routes to enable retrofit of hybrid installations. Speed limits in harbours and channel approaches at beginning and end of daily transits may mean that The Hour Of Power is all that is required for the Battery / Electric part of the cycle.

7.2 THE WORKING DAY FOR WFSV

06.00 to 07.00	1 hour depart dock / harbour
07.00 to 08.00	1 hour open sea transit out
08.00 to 16.00	8 hours in wind farm
16.00 to 17.00	1 hour open sea transit back
17.00 to 18.00	1 hour return harbour / dock
18.00 to 06.00	12 hours alongside

A typical working day starts with the WFSV crew preparing the vessel for departure at 06.00. The vessel would be disconnected from shore power which has been used for overnight battery charging. The vessel can leave the dock and harbour on battery electric power, which also means zero emissions. At the outer sea buoys diesel power takes over for the high speed transit out to the wind farm and to deliver the technicians onto the wind turbines, during this time the batteries are re-charging.

Many vessels have hours of waiting in the wind farms, as they are on standby they cannot anchor. Loitering at low speed and low revs is not an efficient load cycle for diesel engines, plus it leads to increased maintenance and reduced engine life. This loitering period could be on battery power. Later in the day diesel power is used for the high speed transit back to port, the batteries are re-charging. At the outer sea buoys battery electric power takes over to enter the harbour, with zero emissions. Around 18.00 the vessel returns to the dock where shore power is re-connected for overnight charging.

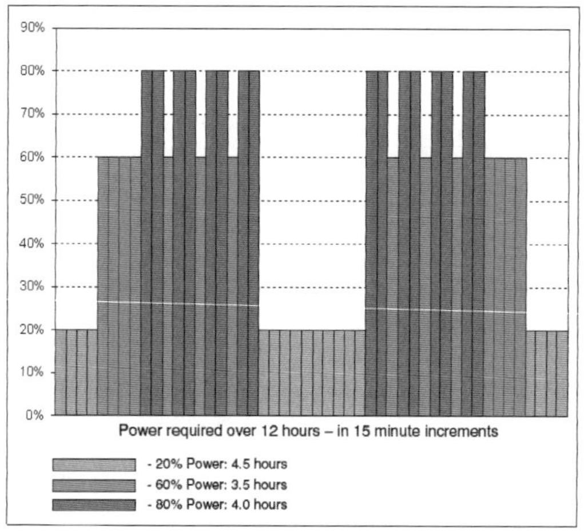

Figure 8: Power required over 12 hours

12 hour WFSV duty cycle utilising Battery / Electric:
Depart harbour: Battery / Electric with zero emissions
At outer sea buoys: Diesels on
Running at speed: Use diesels (charging battery)
Pushing on to turbines: Use diesels (charging battery)
Loitering: Battery power for slow speed or stopped
At outer sea buoys: Diesels off
Return harbour: Battery / Electric with zero emissions
Overnight: Charge battery with shore power

7.3 METRICS & THE HOUR OF POWER

There are various methods to determine the most efficient speed for a vessel to travel at on Battery / Electric power. Naval architects can model this specifically vessel by vessel, but to enable operators to identify approximately how many kilowatts of alternative energy they may need to bring onboard there are simple methods. One of the main criteria that enables Battery / Electric power to work efficiently is to operate the vessel below both hump speed and hull speed. The start of the hump is beyond the point where the boat exceeds its displacement hull speed. To calculate hull speed in knots take the square root of the waterline length and multiply by 1.34. For a WFSV catamaran built to 24 metre (LLL) the waterline in feet will be 80 feet. The calculation is 9 x 1.34 = 12 knots. Running at 20% below this speed is under 10 knots and within many port speed limits.

The hump is clearly defined on the speed / resistance curve for any size of vessel. As speed increases a fuel meter shows fuel consumed, which will significantly increase at the hump. Another source of information is to generate a speed / power curve from the engine management system that shows how many Kilowatts of energy are consumed at a particular speed. For the sake of simple calculation this should be done in zero current and zero wind conditions. Separate allowances can be made for favourable or opposing current and or wind. The objective is identify how many Kilowatts of battery energy are required to replace Kilowatts of diesel energy.

8. NEXT GENERATION BATTERIES

The marine sector can learn from battery advances in aviation and land transport sectors. The next generation of cells and batteries feature key technology developments that are making marine hybrid systems potentially viable. Battery chemistry such as Lithium-ion offer impressive power solutions and the return on investment is starting to fit for commercial operators. Since there is no single system that fits every application it is important to work with manufacturers that have flexibility in cell manufacturing and offer scalable solutions. New factories with fully automated processes are designed to ensure consistently high quality cells and quality control of the entire battery management system.

Lithium-ion battery manufacturers typically offer several variants of High Energy, High Power and Ultra Hi-Power cells which are designed for a wide range of applications. Engineers are developing battery systems that can be integrated with electric and hybrid energy solutions for, transport and grid applications. To date only a few battery OEMs have the experience of taking battery / electric / hybrid projects from concept through cell production into the finished system for maritime applications. Battery OEMs collaborating with boat builders, naval architects and marine operators enables analysis of different workboat duty cycles and gives a holistic approach to development of safe and robust systems. Engine management data can be matched to battery characteristics to develop the most efficient solutions. Onboard energy management systems are designed to ensure that battery systems operate at optimum performance. The objective is hybridizing and electrifying marine vessels to produce commercial benefits and reduced emissions. The design objective is specifying complete systems with power discharge / charge capabilities that are most efficient for their expected use.

Using The Hour Of Power concept as a baseline Lithium-ion battery systems can be designed that consider the footprint, weight, cost benefits and safety case for marine applications such as WFSV. Systems should be modular enabling them to be scaled to suit changing energy levels as required.

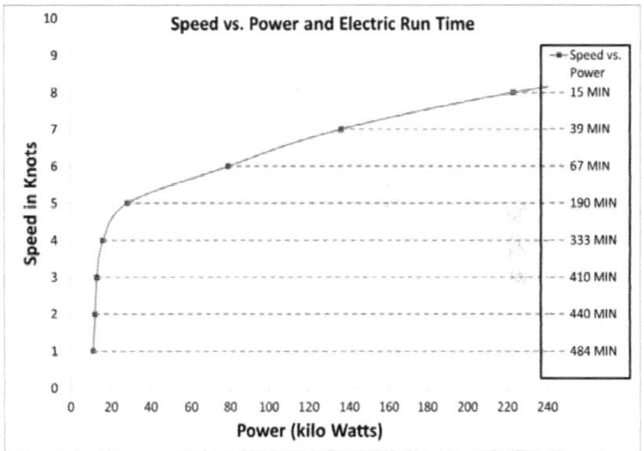

Figure 9: Speed vs Power, 111 kWh System.
Data provided by XALT Energy.

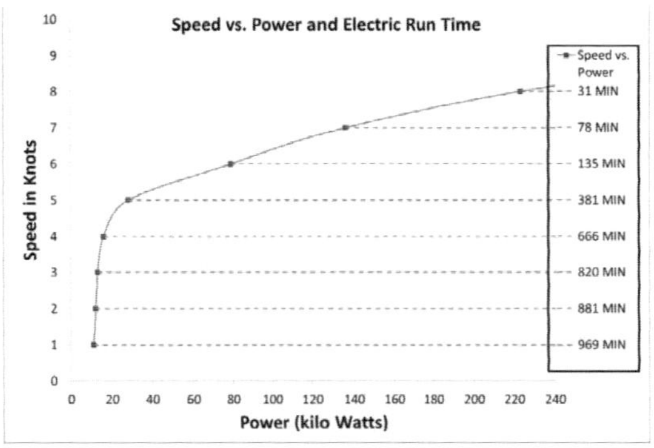

Figure 10: Speed vs Power, 222 kWh System.
Data provided by XALT Energy.

8.1 INSTALLING BATTERIES ONBOARD

Banks of batteries can be permanently installed in each hull on WFSV catamarans. As WFSV have surplus deck space and are designed to carry ISO containers an alternative would be to create portable energy units that could be carried on deck. As many WFSVs are designed to have modular payloads this would enable containers to be moved between vessels as changing duty cycles, or contracts with green credentials, require alternative energy onboard.

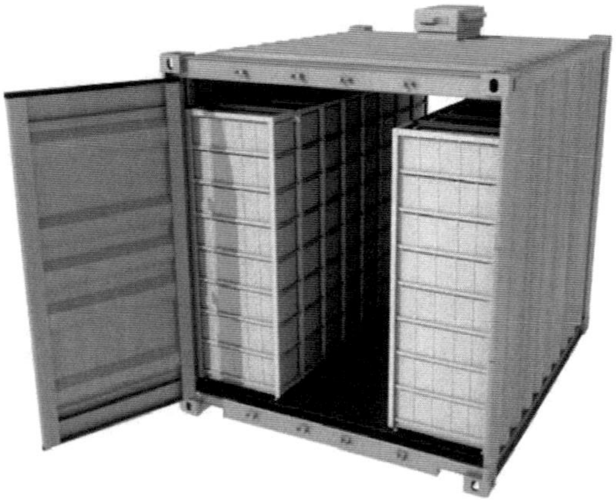

Figure 11: Battery System in 10 foot ISO Container. Credit XALT Energy

9. GLOBAL INTEGRATION

The technology already exists to create state of the art hybrid systems. If the marine industry wants to move forward quickly it will have to build teaming agreements and partnerships that bring together technologies. Hybrid marine systems will not tolerate inferior parts. All components will need to be built to the highest industry standards and designed to integrate globally across multiple platforms. Software architecture will need to be consistent. Where possible, vessels will want a 'plug & play' capability to enable fast replacement of components or technology upgrades.

New high tech companies entering the 'hybrid' market will want to utilise the existing relationships that diesel engine OEMs have with end-users, boat builders and standards agencies. The importance of an international service, spares and support network will add confidence to procurement and life cycle maintenance decisions.

9.1 SCALABLE & INCREMENTAL POWER

Hybrid systems are infinitely scalable which enables end-users to specify what they are trying to achieve over a period of time or an entire fleet. Once a system is defined, projections of savings can be based on engine management data linked to work cycles and applications.

9.2 HYBRID SYSTEMS & COMPONENTS

From over ten years of experience supplying hybrid technology to bus and transport sectors BAE Systems HybriDrive have identified that cost-benefit calculations can be based on both the life cycle of the vessel and the life cycle of the hybrid power system. The 19 metre (62 feet) catamaran research vessel 'Spirit Of The Sound' runs virtually silently on battery electric power utilising BAE Systems HybriDrive system for two hour study cruises on Long Island Sound.

Figure 12: The research vessel 'Spirit of the Sound'. Equipped with BAE Systems HybriGen technology.

Siemens have extensive experience of hybrid and electric technology for various modes of transport. Siemens offer both series and parallel hybrid systems for the commercial marine market. Components related to the electrical propulsion system include the drives, generators, inverters, filters, and control system, plus step up gearboxes to attach the generator to diesel motors.

Figure 13: 65 metre Italian Coast Guard vessel with hybrid propulsion system. Credit Siemens AG.

ZF Marine offers a range of hybrid-ready transmissions and for larger fast craft applications. The design is based on a unique Power Take In (PTI) configuration, allowing highest flexibility for customizing installations. These transmissions can be integrated into hybrid propulsion systems for all types of fast craft, from coast guard vessels to fast offshore supply vessels.

9.3 SERIAL PRODUCTION & SMALL CRAFT

German electric outboard motor manufacturer Torqeedo has developed high-power integrated propulsion systems for serial production. Torqeedo Deep Blue hybrid power and propulsion for the professional marine sector is based around components of the proven and multiple innovation system for electric hybrid yachts. The hybrid system can provide drive systems and supply the power required onboard. The use of renewable sources of energy, including solar power, has also been integrated into the energy management system. Professional users of the hybrid system can travel silently and use environmentally friendly sources of energy in port and at lower operating speeds. This is currently only available for lower KW ranges but as the technology is proven the components can be scaled up.

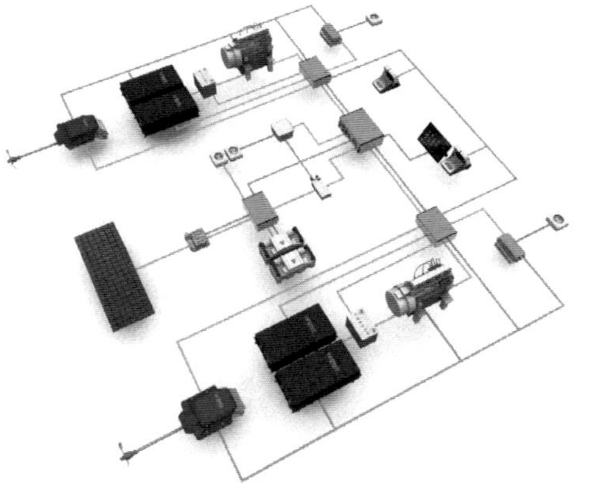

Figure 14: Deep Blue Hybrid by Torqeedo.
A fully integrated hybrid power & propulsion system.

10. LEGISLATION & STANDARDS

Various legislation and standards have implications for end-user organisations, boat builders and equipment manufacturers. As cells, batteries, power generation and storage evolve it is important to address misconceptions and myths to enable progress. For example Lithium Ion has been part of our lives in non-marine sectors for many years as the battery power behind electric cars, city busses, smart phones, tablets, laptops and cameras.

Besides how to present performance metrics, it will be essential to identify whether current standards from land based applications are valid. An environment with electricity, water and damp enclosed spaces requires specific testing and sea trial standards. New international standards for high voltage propulsion systems and comprehensive energy management in maritime environments will be relevant.

As technology is moving at a faster pace than regulations it is important that guidance notes need to be non-prescriptive and linked to state of the art knowledge as it develops.

Lloyds Register has issued a 'Battery Installations Guidance Note' relating to key hazards to consider and their approach to approval. DNV-GL has issued a 'Guideline For Large Battery Systems' and a guide titled 'The Future Is Hybrid'.

In 2016 the UK Maritime Coastguard Agency (MCA) released Maritime Guidance Note MGN 550, 'For Safe Design, Installation and Operation of Lithium-ion Batteries'. The guidance, for ship designer builders and operators, is intended to facilitate safe and environmentally-friendly Lithium-ion battery solutions for vessels utilising lithium-ion batteries as part of a hybrid power system or as the sole source of propulsion power.

11. BACK UP SYSTEMS & REDUNDANCY

Power and propulsion systems that are designed and built for professional or commercial operations need to run hard, often for long hours in adverse sea conditions. Professional boat operators around the world have learned that power and performance are relevant, but reliability and durability are important factors for all types of engine and propulsion. Users must be able to rely on these systems at all times. In certain situations failure is not an option – the engineering must not break, but if it does there needs to be an instant back-up.

Hybrid systems can offer redundancy. A WFSV with Battery / Electric capability linked to x2 main engines, x1 generator and battery bank has various power options:
x2 Main Engines: Directly powering propulsion
x1 Main Engine: Directly powering propulsion
x1 Generator: Running electric powered propulsion
Battery Bank: Running electric powered propulsion

12. WFSV USING RENEWABLE ENERGY

The IMO Energy Efficiency Design Index (EEDI) aims to reduce emissions from large ships by 30% between 2014 and 2025. Countries developing new green energy policies may want to enter the offshore wind, tidal and wave energy sectors by specifying vessels that harmonise with the global reduction in use of fossil fuels.

Indentifying the viability of hybrid diesel / electric power for offshore wind farm support vessels is an interesting project that links green energy onboard with renewable energy from the environment. Hooking up to offshore wind farm turbines may even provide charging options.

Many wind farms are 'overplanted', this means there is more energy generated by wind than is required for the grid on land. This surplus allows for downtime and maintenance of turbines. For offshore wind farms in sea areas such as the Dogger Bank in the North Sea which is over 80 miles from land, wind generated electricity could be utilised at source from the local offshore grid to power hybrid vessels working in the sea area.

13. LOOKING TO THE HYBRID FUTURE

There are clear opportunities related to hybrid power, however the integration of products will be crucial. There will be a new career in shipyards for 'Hybrid System Integrators'. These will be individuals and teams that have a holistic approach to engineering and many will need to be qualified high voltage technicians. When integrating hybrid systems a shipyard will need to identify which competencies are required. To warranty the system it will need to define who signs off the installation and the components on handover. A single point of failure at sea will reflect on all OEMs in the hybrid power 'system'.

For naval architects class rules, safety, performance and cost are relevant when considering innovative battery power and diesel electric propulsion systems. The challenge for designers is to engineer solutions utilising hybrid technology which are affordable plus manageable in terms of physical size, weight and maintenance.

Cost / benefit implications will start with the initial purchase of the system then work out payback periods based on the life cycle of the vessel and life cycle of the hybrid power system. Once a system is defined projections and audits can be based on engine management data linked to work cycles.

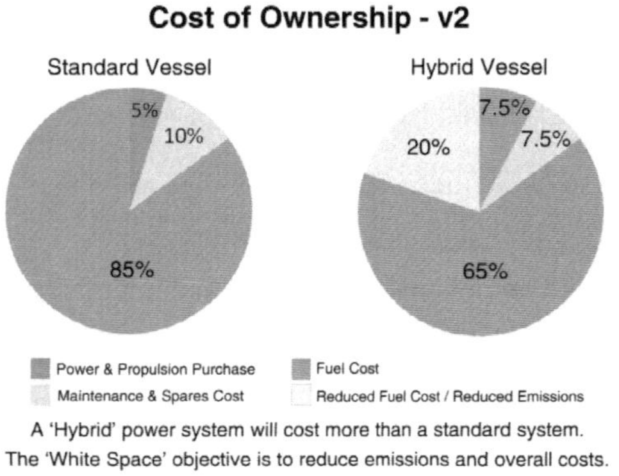

Figure 15: Example of how the cost of ownership for a power and propulsion system could be compared

The engineering required for a hybrid power and propulsion system involves higher capital outlay than a conventional system. To overcome this financial resistance in the automobile market, various countries have applied subsidies to the purchase cost and reduced vehicle taxes for commercial electric and hybrid vehicle operators. This approach could work to accelerate adoption in the maritime sector. Hybrid systems are infinitely scalable which enables owners to specify what they are trying to achieve over a period of time or an entire fleet. Objectives include reduced fuel costs, reduced maintenance costs, extended engine life and reduced emissions.

14. CONCLUSIONS

As new sources of energy become available it is important to identify which energy source best fits the vessel, duty cycle and environment to give efficient power when it is needed. Since no two vessels, routes or captains are alike, decisions can be improved with data logging and analysis.

Two recent developments caused operators, owners and builders of WFSV vessels to consider hybrid marine power. Firstly new emissions laws in ports and secondly subsidies applied to offshore wind farms in UK waters are likely to reduce. For the North Sea wind farm industry to move forward in a genuinely sustainable manner it needs to utilise expertise from other sectors to drive innovation, improve efficiency and support relevant safety standards.

15. ACKNOWLEDGEMENTS

The author would like to express great appreciation: To BAE Systems HybriDrive, South Boats IOW / Alicat, XALT Energy, Siemens, Torqeedo, Hybrid Marine for background information and images. To Dr John Warner, Robert Young, Bernard Twomey and Tania Berry of Lloyds Register, Dr Davion Hill and Dr Ben Gully of DNV-GL for information on batteries & safety standards. To UK MCA and USCG for exchange of knowledge.

16. REFERENCES

1. 'Hybrid Marine Power & Propulsion', Conference - Poole UK, 2015

2. 'Design & Construction of Wind Farm Support Vessels', Conference - London UK, 2016

3. 'Next Generation Marine Power & Propulsion' Conference - Southampton UK, 2016

17. AUTHORS BIOGRAPHY

John Haynes holds the current position of Managing Director at Shock Mitigation Ltd. Maritime expertise includes 30 years training, consultancy and strategic product development focussed on the professional sub IMO / sub 24 metre sector. He is an Associate Fellow of The Nautical Institute, Associate Member of RINA, Yachtmaster Ocean and Advanced Powerboat Instructor. He has organised and chaired conferences including: Hybrid Marine Power & Propulsion (2015) / Next Generation Marine Power & Propulsion (2016). In 2015 he brought a working group together to assist UK MCA develop MGN 550, the Maritime Guidance Note: 'For Safe Design, Installation and Operation of Lithium-ion Batteries'. He has written specialist articles on future requirements and new technology for over 20 international publications. At Seawork 2016 he was voted Maritime Professional of the Year.

DEVELOPING A HOLISTIC STRATEGY FOR SHOCK MITIGATION ON FAST BOATS

J Haynes, Shock Mitigation Ltd, UK

SUMMARY

The definition of shock mitigation is, 'to make a violent collision or impact less intense'. A shock mitigation strategy is essential for all vessels that undertake open sea transits or operate in rough water. A challenge for the builders of fast workboats and wind farm support vessels is delivering platforms that balance performance with the physical demands on crew and passengers. The consistent objective is that passengers, including technicians, arrive safely at their destination ready to perform a task.

CAD and digital modelling are key components in the designing of high speed craft, but feedback from the human body is crucial input that designers and naval architects must consider for next generation vessels. Since the 1990's, focus has mainly been on developing mechanical suspension seats to reduce the effects of vertical accelerations. Areas requiring further investigation include fore-aft, lateral and vector forces, plus improving seat cushion performance and comfort.

NOMENCLATURE

RS: Repetitive Shock
WBV: Whole Body Vibration
WFSV: Wind Farm Service Vessel
CTV: Crew Transfer Vessel
RIB: Rigid Inflatable Boat
RIB: Rigid Hull Inflatable Boat
RNLI: Royal National Lifeboat Institution
IMO: International Maritime Organisation
LLL: Load Line Length
MMI: Man Machine Interface
RAG: Red - Amber - Green
PPE: Personal Protective Equipment
COTS: Commercial Off The Shelf
FLIR: Forward Looking Infra Red
IT: Ischial Tuberosity
MRAP: Mine Resistant Ambush Protected
MCA: Maritime Coastguard Agency

1. INTRODUCTION

A force that all crews of planing craft are aware of at sea is when a vessel climbs over the top of a wave and occupants experience a negative gravity sensation, similar to the humps on a roller coaster. As the boat lands the Vee of the hull displaces water, then due to buoyancy it can go no deeper. A few milliseconds later the occupant's bodies decelerate, with the impact felt through the legs and feet if standing or the backside if sitting. The effect of these high G forces on the human body can cause fatigue, plus acute or chronic injury.

Medical doctors and physical therapists can predict the weak points of the human skeleton, along with discs and muscles, from this type of activity. Groups of maritime professionals, when asked where they experience pain during fast craft transits in wave conditions, will typically list lower back and knees in that order of incidence. Other reported injuries include neck, upper back, hips and ankles.

2. DEVELOPING A SHOCK MITIGATION STRATEGY

Planing craft are selected to perform a wide range of operations. Professional organisations using fast boats need to identify what level of sea conditions are likely to be encountered, then ensure that the type and size of craft specified are fit for purpose.

The consistent objective is that commercial passengers arrive safely at their destination ready to perform a task. For wind farm technicians the task is typically to step onto a ladder then climb a turbine tower and carry out physical tasks in an exposed location for a number of hours each day.

Maritime pilots need to arrive alongside of a ship that is underway then climb a ladder. They then have to use a high level of cognitive ability in the ship's bridge. Maritime search & rescue teams need to arrive on the scene as quickly as possible, make rapid decisions, then be able to physically lift casualties from the water.

Military passengers may need to arrive 'fit to fight'. For some navies increasing individual fitness and stamina is seen as the solution. In other parts of the world people are deemed to be expendable as there are others are ready to take their place, but that approach does not help when a boat and crew are underway with a task to complete.

Technical shock mitigation solutions currently include more efficient hull forms, appendages to improve trim and ride quality plus suspension seating and cushion materials. With an effective shock mitigation strategy the helmsman, crew and passengers benefit from increased comfort and reduced injury.

But shock mitigation is not just about reducing injury. An organisation can increase sea time for assets, cover greater distances at higher speeds, improve crew performance and extend operational effectiveness.

3. LEARNING FROM OTHER TRANSPORT SECTORS

It is important to learn from other sectors that have made progress with shock mitigation, but myths need to be dispelled. From motion analysis metrics on land, sea and air craft it is clear that not all vehicle impacts are simply lesser or greater G forces. A major difference between automobiles and boats is the suspension system managing vibration between road wheels and the chassis. Trucks take this a stage further and have suspended cabs, therefore the seat is mainly for comfort. An agricultural or mining vehicle driving over rough terrain experiences different loads to a boat at planing speed on rough water.

In recent years the development of MRAP (Mine Resistant Ambush Protected) military vehicles has accelerated research into reducing the effects of mine blast. However a vehicle seat that mitigates the shock from a single mine blast is unlikely to mitigate the repeated slamming effect from wave impacts on high speed craft.

Aviation has researched shock mitigation and the effect of various impacts. Helicopter seating is designed to protect the crew from hard landings and a crashworthy seat is part of the overall crumple zone. Ejector seats have saved the lives of many pilots and navigators as they exit from fixed wing aircraft. However ejection is a single event based on a consistent input force, usually initiated by the seat occupant, and compared to loss of life some level of injury may be acceptable.

Large wave slams at sea are not usually isolated events, they can be of random magnitude and from multiple directions. In rough sea transits the marine suspension seat has a fraction of a second to return from a 'hit' to mitigate a 'double hit' or the next pattern of multiple impacts.

4. THE EVOLUTION OF FAST BOATS

Fast motor patrol boats were used in the Second World War and during the Vietnam conflict. The RIB (Rigid Inflatable Boat) concept originated from Atlantic College (Wales UK) in 1963 and was adopted as a fast rescue craft by the RNLI (Royal National Lifeboat Institution).

In the 1960s and 1970s GRP / FRP speed boats were generally used for leisure sports such as cruising, waterskiing or racing.

As requirements changed during the 1980s and 1990s, professional sector RIBs increased in waterline length and capability. On fast craft between 7 metre (23 feet) and 11 metre (36 feet) the number of occupants typically ranges between 6 and 12.

Since the 1990s RIBs have been used for offshore racing, including record crossings and circumnavigations. Slender hull forms help to reduce shock at high speeds in waves and race boat designers, with no payload requirement apart from fuel, can position their crew in the best position to minimise injury to occupants.

In the 1980s criminal organisations capable of acquiring fast boats were most likely running contraband or illegal migrants in from offshore. As boats were purchased the simple questions 'what payload' and 'how fast' were being asked from the Mediterranean to the Caribbean. These organisations soon owned the fastest boats on the water. Seating was irrelevant and crews were usually standing on hard decks as any wasted space or weight reduced the all important payload.

Identifying a boat capable of 50 knots was an important criteria, as one hour transits from under 50 miles offshore were often too short for enforcement officers to launch their vessels and intercept.

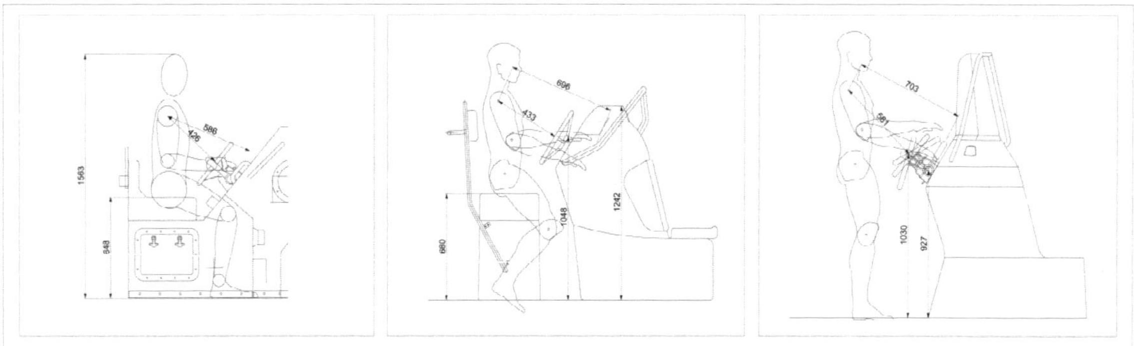

Figure 1:
Low Jockey Helm Seat
High Jockey Helm Seat
Standing at Helm
Image Credit: Kobus Design / John Haynes

5. GO FAST OPERATIONS AND SHOCK MITIGATION

In the early days of counter 'go fast' operations for marine police, customs and borders organisations the focus was firmly placed on procuring high performance hulls to chase down offenders. Boat crews and enforcement officers were usually standing in wrap around bolsters or leaning against padded rails. There were often no seats and even when fitted these physically capable operators were more likely to stand using their experience, balance and leg muscles to ride the boat.

Countering a high value, high speed run in from offshore required a capable fast response boat with physically fit crews ready for a short sharp pursuit, usually at night, that often ended in a 'lights off' high speed chase in the inter coastal waterways. This was a brief one on one, win or lose, situation with lucrative incentives for a single evasion. The speeds achieved by 'cigarette boats', named after their long narrow profile, on weekend offshore powerboat races in the US suddenly became significant to the maritime law enforcement community.

The history of 'shock mitigation' technology can be traced back to the 1990's and the rapidly changing requirements of counter narcotics and military marine operations as boats became faster with extended and more complex missions. However the significant event for homeland security in port and maritime security operations was 9/11. As the global stakes changed and requirements were analysed it became obvious that security and military boats would need to operate faster than ever before. The people on them would need to be 'fit to fight' determined adversaries after a fast transit in extreme sea conditions.

6. SUSPENSION SEATING AND THE MAN MACHINE INTERFACE

As high speed craft designers and professional sector boat builders rose to the challenge they built in 'over engineering' wherever possible. In the early days of shock mitigation 'feet off the deck' suspension seating incorporated complex engineered solutions, often adapted from land vehicles. Other manufacturers then started to offer lighter and simpler 'feet on the deck' jockey suspension seats.

Next generation manufacturers have created a range of suspension seats and adjustable shock mitigation systems to enable humans to operate fast craft in the roughest of conditions. These solutions utilise sophisticated mechanical engineering to disconnect the boat occupant from G forces as the hull moves over rough water and wave conditions. Taking this a stage further, the latest shock mitigation technologies now utilise Man Machine Interface (MMI) to link coxswains, navigators, engineers, FLIR operators, gunners, communications operators and commanders to their tasks, controls and screens. These designs need to be integrated into the general arrangement of the deck area or cabin at design stage and most cannot be retrofitted into existing craft due to structural limitations and console heights.

7. SHOCK MITIGATION BY ALL MEANS

With increasing reliance on high technology there is an opportunity to step back and define what and where are the forces that we are trying to reduce on the water. It is also important to be conscious that for boat occupants at sea on a dark and stormy night 'simple is good'.

Shock mitigation is important for people who want to go further, faster or for longer in extreme conditions. One of the things that marine professionals have in common is a growing awareness of injury. As crews need to focus on the task at hand it is important to protect all occupants of fast boats from injury, whether they are sitting or standing. RIBs and high speed craft are an efficient means to get afloat and travel fast, but as technology changes even a very fit human body has its limits. When a RIB is running in flat conditions standing occupants experience vibration coming up through their feet on the deck, held there by gravity and body mass. Lightweight, agile people in flexible footwear and clothing have a physical advantage over heavyweight military personnel wearing boots and heavy body armour.

In flat sea conditions there is vibration from the engine but boat occupants are not exposed to harm from this level of vibration. Fast boat operators know that waves from groundswell, wind and even other vessels wakes can change everything. As powerboats get lighter and faster there are new issues to consider. Operating a fast boat in flat open waters is relatively easy. But operators running powerboats at high speed need to understand the forces they are dealing with. Vibration and impact affect novice crews and less experienced passengers when the boat hits a wave as they do not know how to brace against the forces that experienced crews react to intuitively.

Professional high speed craft operators need a combination of shock mitigation solutions. To date there is no 'silver bullet', shock mitigation is about reducing forces and potential injury by a few percent wherever possible via the hull, seats and deck. It is agreed that the overall effect of reducing the effects of peak acceleration will be reduced incidence of injury.

8. RS & WBV - REPETITIVE SHOCK AND WHOLE BODY VIBRATION

Whole Body Vibration (WBV) exposure on planing craft is usually caused by continuous 'hammering' from short steep seas or wind against tide conditions. Repetitive Shock (RS) on planing craft is usually caused by random 'hits' from head sea impacts, crossing seas or overtaking following seas.

The EC Vibration Directive came into force in 2010 and there is now an increasing awareness that professional operators need to reduce the effects of RS and WBV on their employees by all means. Fast boat drivers know that even moderate sea conditions can be painful and may lead to serious injuries to lower back, knees, ankles and neck. What they often do not realise is that a fast boat in 1 metre (3 feet) seas can measure impacts of 5g to 10g on the deck.

At commercial boat shows around the world it currently appears that the fast craft sector has adopted suspension seats as the main shock mitigation solution to counter effects of RS and WBV. In reality the new craft exhibited only represent a small percentage of vessels. The majority of in-service craft belong to hard working organisations with budget constraints and limited space on deck for larger seats, or craft which are close to their payload for lifting by dockside crane or onboard davit.

9. UNDERSTANDING AND QUANTIFYING THE FORCES

To develop the next generation of shock mitigation solutions for fast boats operating in waves it is important to understand the forces that the hull is experiencing as it passes through or over moving and uneven water surfaces. Disturbances could be caused by storms hundreds of miles away generating groundswell, local wind blown waves or the wake of other vessels even in calm conditions.

From an understanding of wave events scientists and engineers can analyse the exposure chain of hull, deck, seat and finally the human body of boat occupants.

Scientists are starting to quantify wave slams and describe how differently a planing craft re-enters the water in various sea states and wave patterns. In simple terms a fast RIB or high speed craft can land bow first, stern first, flat on the keel, or on one side of the hull. The process can be filmed and replayed at slow speed or analysed with data recorders in different sea states. When crews come off the water after a hard transit the subjective statement of 'that was a hard ride' can now be quantified. Two coxswains on identical boats running side by side at the same speed can have vibration measured on each craft for later analysis. Used correctly this approach is a powerful training tool where video and metrics can show 'what good looks like'.

10. DATA LOGGERS AND DECISION MAKING

Monitoring the effects of vibration and impact on the hull or critical pieces of equipment is relatively simple as data loggers can be attached to GRP / FRP, aluminium or plastic in various ways. Land based industry has been doing this for many years on factory and moving machinery, then developing shock mitigation methods to dampen or eliminate vibration by tuning the suspension system. Measuring the forces on the deck or on a seat base will produce data, but a topic of major debate in recent years has been how to gather vibration data from the crew and passengers. Lower back is the area with highest incidence of injury, however fixing accelerometers to a flexible human body is difficult. Seat pads and kidney belts are common methods that provide reasonable contact for data loggers. The consistent objective is valid and reliable measurement for the assessment of RS and WBV.

Figure 2:
Round Britain record holder 'Hot Lemon' RIB
With crew standing – supported by bolsters.
Image credit: Mike Deacon

As the cost of data loggers has come down various organisations around the world are installing these on in-service craft to gather and store vast amounts of vibration data. Analysing the data and deciding which metrics are relevant is a challenge. A simple Red - Amber - Green (RAG) index showing traffic lights on the console is an indicator to show crews that vibration exposure is increasing. However setting the vibration thresholds for colour change may vary depending on which measuring method is used and whether a crew will be allowed to ignore the warning signals during a critical mission.

11. SITTING AND STANDING ON FAST BOATS

Research in the past decade has led to a better understanding of the forces affecting the human body at sea. Professional organisations are now asking what they can do to reduce the effects of vibration and potential injury as fast boat transits take their toll on the fitness and long-term health of personnel.

On jockey or straddle style suspension seats the occupants feet are still on the deck. In a head sea or following sea transit, some of the forces are vertical (usually referred to as Z axis) and some of the forces are fore and aft (usually referred to as X axis). The boat rarely moves sideways (usually referred to as Y axis) but side-on seas cause random events and a combination of forces that cannot be anticipated which create the most damaging vector of forces on the human body.

Jockey or straddle seats enable occupants to sit or stand in the same space on the vessel. In a sitting position, with legs bent at close to a right angle, the majority of bodyweight is on the seat cushion and the legs are used for balance. In a standing position all the bodyweight is on the deck, with knees and thighs used to grip the seat for balance.

A leaning position is where the seat occupant stands, locks their legs out in front of them and uses the seat backrest as a leaning post. Locked legs limit the use of knees, ankles and muscles to absorb impacts. It is important that all crews and passengers understand basic biomechanics, including the risks of a poorly aligned or twisted spine. Designers with a good understanding of ergonomics should correctly position hand holds and wheel / throttle controls to encourage good posture.

On certain craft there is the option to retrofit mechanical suspension seating. However introducing high suspension seating on vessels with low consoles and work stations is not effective and may cause poor ergonomics that increase the risk of injury. If seats are too large to fit the space, too heavy, or too expensive for the organisations budget, then operators will need to find other methods to reduce the effects of vibration and impact on the human body.

With boats now expected to last 15 to 20 years many military, police, government and commercial organisations will not retrofit suspension seats and crews will simply have to continue to stand in anything other than calm conditions. Military and professional operators around the world are now looking at the deck and considering how the forces could be reduced there.

12. STANDING ON BARE DECKS AND CUSHION MATERIALS

On boats without suspension seats experienced operators usually choose to drive RIBs standing up in anything but flat conditions. As the seas get bigger most fast boat operators prefer to stand to see oncoming waves and use their legs to balance and absorb the boats movement. At that point they also have the opportunity to engage and use an incredibly complex suspension system - the human body.

Figure 3:
Professional RIB Helm Station.
Helmsman Sitting / Leaning / Standing.
Image Credit: John Haynes

To enhance the standing position crews need to consider what they are standing on. Typical decks made of aluminium, fibreglass and composite materials transmit shocks and vibration. Standing on foam or rubber flooring is an option, but most of these materials lack stability which causes fatigue. This is where the constant 'firing' of ankle and calf muscles eventually leads to shaking legs. When simply standing on most foam products the material is already pre-loaded, which means that the feet sink towards the deck, therefore reducing the depth of cushioning. For heavier occupants in wave conditions it becomes increasingly likely to bottom out, which feels like the underside of the feet have been hit with a hammer.

13. IMPACT MITIGATING BOAT DECKING

Impact mitigating decking can be particularly beneficial on boats with leaning posts and on boats with fixed jockey or straddle seats where the occupants stand in rough conditions. Even 'feet on the deck' suspension seats require the occupant to use their leg muscles and carry a percentage of body weight on the deck, this can result in a significant weight on one leg particularly in crossing or side on seas.

The shock transfer to crews and passengers standing on non-cushioned deck surfaces is significant when operating at speed, even in moderate sea states. Just as suspension seats are more engineered than standard foam seats, impact mitigating boat decking solutions need to be more engineered than simple rubber or foam.

SKYDEX Technologies have developed high-performance boat decking products that use patented energy-absorbing geometries to reduce the effects of impacts and dampen vibration on RIBs and high speed craft. The objective is to reduce the potential for injury to personnel, and protect equipment carried on deck, when exposed to wave slam events in fast craft operations.

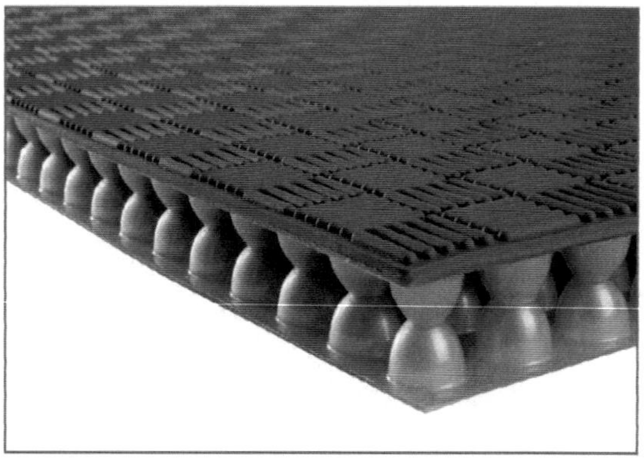

Figure 4: SKYDEX Shock Mitigating Boat Decking utilising elastomers & geometries

SKYDEX Boat Deck is available from 14mm to 50mm thickness, based on the level of deck coverage and mitigation required. This can be fitted on new boats or retro-fitted to in-service vessels to reduce the effects of shock and vibration. The SKYDEX technology that has been utilised for marine decking is the same as the blast mitigation flooring used to protect the occupants in over 20,000 military vehicles to date. The technology works by utilizing elastomers and proprietary geometries to create a multiphase spring that is engineered to react differently to various levels of impact. SKYDEX works differently than foam and is tuned to react to the expected levels of shock. In a wave impact situation, particularly in the dark, the human body cannot react fast enough to protect itself from the shock. SKYDEX reacts instantly to the forces, plus the technology works in all axes to reduce the effect on the human body whether the boat is landing vertically, into head seas or even side on.

Figure 5: SKYDEX Boat Decking, on RIB with shock mitigation seating

14. SEATING SOLUTIONS EVOLVE

As the fast boat sector has evolved over the past decade the diversity of potential seating solutions has grown with it. RIB and high speed craft seats need to provide a good ergonomic position for the different requirements of helmsman, crew and passengers. Comfort is important therefore seating design needs to allow space for occupants PPE (Personal Protective Equipment) and floatation devices. Some organisations also require webbing and body armour to be worn and weapons or tools to be carried. This may require an adjustable seat with built in personal storage for equipment.

For the past thirty years typical seating for professional RIBs and fast craft has been jockey (straddle) or leaning post with foam cushioning. Due to higher operational speeds many organisations are now using or considering suspension seating. The objective of a suspension seat is to separate the helmsman and crew from the worst effects of vibration and impact.

Suspension seating may have feet on or off the deck. A jockey suspension seat is where a percentage of the occupant's weight is spread between backside on the seat and feet on the deck. Typical stroke or travel of the suspension is 150mm (6") to prevent bottoming out. The next generation of seats have the option to adjust for height and weight which allows for a broad size range of occupants. This adjustment has to be supported with clear instructions and onboard briefing as a badly adjusted seat can have a negative effect on the occupant.

A conventional suspension seat, with feet on attached footrests, carries all of the occupant's weight. These are usually in wheelhouse or cabin boats. Suspension travel for semi-displacement craft seating starts around 100mm (4") and for extreme fast craft the travel on feet off the deck seats can be over 200mm (8"). It is becoming increasingly relevant to either select the correct seat for the boat or tune the seat to the boat and its expected use.

15. SPEED & THE BALANCE OF POWER

The traditional operating model for shipping is moving goods from point to point as fast as possible. When 'slow steaming' is introduced the vessels speed is slowed down in order to lower costs by reducing fuel consumption. This also changes vessel motions for occupants.

The requirement for 'power' is changing for many professional fast craft operations. Organisations are less focussed on overall speed and are now considering other factors during transits including the long term health of crews and the Man Machine Interface (MMI). Many onboard systems and electronic devices cannot be operated at high speed and it is not acceptable to injure personnel during sea transits. In the offshore sector technicians need to be fit to climb a wind farm tower to do manual work at the end of a transit.

16. WIND FARM SUPPORT VESSELS

Current fleets of Wind Farm Support Vessels (WFSVs) have been specifically designed to work in the wind farm support sector. As catamarans offer reduced resistance to motion and reasonable stability they are now the hull form of choice for most WFSV working in northern European waters. As the demand for wind farm support has intensified, vessels have been specifically developed for transiting to wind farms in a wide range of conditions then delivering technicians onto wind farm turbines.

In the early stages of wind farm development vessels ranged from 15 to 18 metres (50 to 60 feet). Due to the International Load Line convention vessels are typically delivered below 24 metres LLL (Load Line Length). In recent years the majority of vessels have been designed to 24 metres with wide beams to get maximum working space and payload onboard. When moving technicians transit speeds range between 20 and 25 knots with some vessels having a top speed of 30 knots.

Figure 6: 24m WFSV operated by Seacat Services. Credit: Alicat / South Boats IOW

16.1 TRANSITS & TURBINE TRANSFERS

Transit time and accessibility to the wind farm are often weather dependent. In the North Sea, sites further from shore typically experience larger significant wave heights. Transferring maintenance personnel and their equipment from vessel to wind turbine safely in various sea states is a major challenge in offshore wind farm operations. In these locations the turbine foundation is fixed to the seabed resulting in considerable relative movement between the vessel and the structure.

Turbine transfers are where a vessel uses engine power to push the bow against 'j-tubes' which run vertically on the outside of the turbine access ladder. Thrust is applied to increase friction and reduce motion between specially designed bow fenders and the landing platform. Maintenance technicians step onto the turbine access ladder which is set back from the j-tubes providing a safety zone to prevent personnel on the ladder from being crushed. Large waves, especially when coupled with strong currents, can cause the vessel to lose position or contact with the turbine.

Figure 7: 26m WFSV operated by Seacat Services. Credit: Alicat / South Boats IOW

17. SITTING AND RECUMBENT SEATING FOR CREW TRANSFER VESSELS

The European offshore wind farm sector is driving higher levels of safety for crews and technicians being transported on fast boats. This is a demanding industry with hard deadlines where technicians spend as many working days at sea as weather will allow and working days are as long as possible.

Crew Transfer Vessels (CTVs) are passenger ferries to and from the wind farm area and between individual wind turbines. Crash testing for individual seats and deck fixings is relevant for craft operating at planing speed in close proximity to turbine towers and other vessels, particularly at night or in poor visibility. Improved ergonomics and adjustments will be required for these seats as passengers are expected to travel further offshore in higher sea states for longer periods. Varying body position during a long journey is important on any mode of transport.

When the seat occupant is asleep the whole body needs to be suspended to enable the suspension to work automatically. Ongoing research is considering whether a recumbent or reclining position is more comfortable and gives better spine protection when conscious bracing is not possible. In rough conditions the seat is the only part of the vessel that humans can travel on so increasingly workstations will be attached to the seat.

18. SUSPENSION SEAT TESTING

To support procurement decisions the next generation of fast craft hulls and shock mitigation solutions need to be compared. Full scale sea keeping trials in high sea-states are costly, weather dependent and will vary between craft. Wave conditions are rarely consistent over sea areas or over time. It is also hard to justify taking people to the point of injury as part of testing.

Shock mitigating seating has the potential to improve the environment on board a high speed craft for operators and a large number of seating technologies are available 'off the shelf'. However, to help select equipment that is appropriate for the vessel and its intended mission it is necessary to test such seats under controlled, repeatable, conditions to expose them all to the same excitation.

To address this, an International Standards Organization (ISO) Technical Committee Working Group was established in 2013 to develop standard procedures for 'Laboratory Evaluation of Marine Seat Shock Isolation'.

Preliminary drop testing has been performed in the UK and Canada to demonstrate shock mitigation characteristics of various seat types in the lab before installation on boats for on water testing. Procurement decisions for seating on new fleets of boats are now being made using the combination of lab test data and subjective feedback from experienced crews.

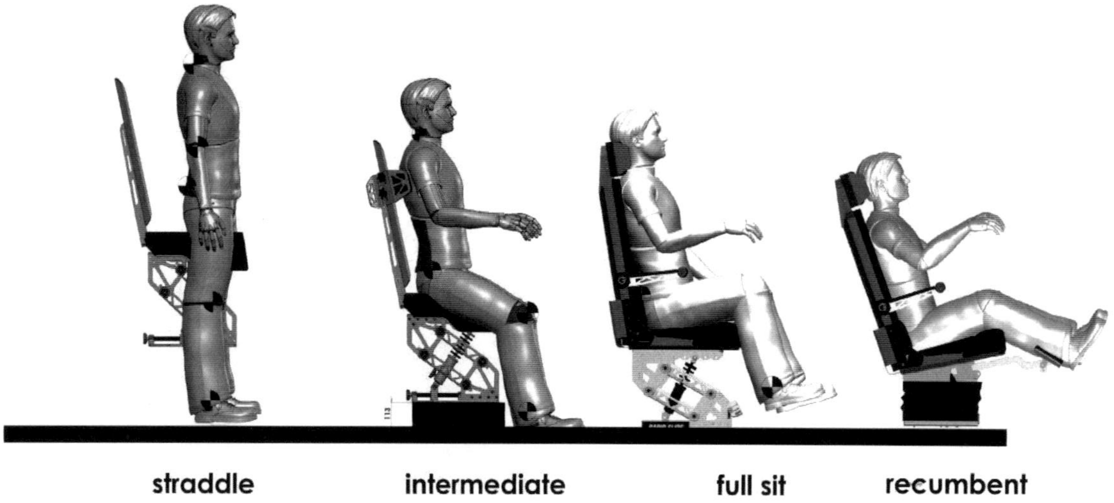

Figure 8:
Suspension seat positions range from semi-standing to recumbent.
Straddle: standing position has all occupant weight on deck.
Intermediate: has occupant weight distributed between seat and deck.
Full Sit: has all occupant weight off deck in upright position.
Recumbent: has all occupant weight off deck in reclining position.
Image credit: KPM Marine & Scot Seat

19. NEXT GENERATION VESSELS AND THE MAN MACHINE INTERFACE

Foot straps are installed on RIBs and high speed craft for various reasons. These may be the only means of staying connected to the vessel if hands are occupied on tasks requiring fine motor skills such as controlling fly by wire controls, navigation systems and computers. In combat situations, turns and emergency stops can be so extreme that both hands and feet are required to keep occupants in the boat. Some suspension seats create a potential ejection moment as the damper and spring rapidly return the suspension from a compressed position. Foot straps benefit from improved design to enable simple adjustment of the opening plus a range of fore and aft positions. This allows for different heights of seat occupants, plus all sizes of people have different positions for their feet whether sitting or standing over a seat.

Linking the man or woman to the task is becoming a requirement for many military and Special Forces organisations. As seat and suspension systems evolve more responsive controls and work station layouts with improved ergonomics are part of the Man Machine Interface (MMI) and human factors design. Helmsmen already have joystick steering and throttle controls built into the armrest of their seat. Navigators can only work effectively at high speeds if their GPS and radar screens are moving in unison with their hands and eyes. Communications, situational awareness and weapons technicians benefit from control panels attached to their seats. Gunners operating at high speeds require specialist suspension seats attached to weapons.

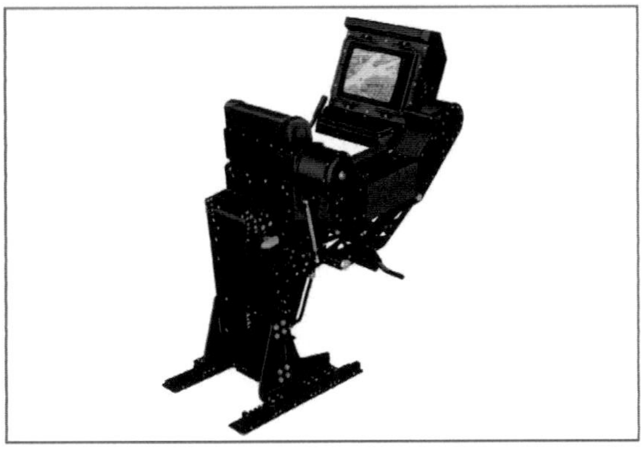

Figure 9: Man Machine Interface (MMI) Suspension seat and workstation combined. Image credit: SHOXS

In recent years there have been significant developments in the fast boat sector. Major steps forward in technology by seat manufacturers mean that 'man' is not always the weakest link on fast boats. Boat speed on flat water is governed by the engine power limit, but speed in waves may be governed by structural design limit of the hull. Professional fast boat coxswains generally operate to a level known as tolerable discomfort, then the pain threshold is reached and they are likely to slow down. But as innovative suspension systems have de-coupled the man from the boat the new challenge is to build hull, components and equipment that are able to survive the extreme ride severity that occupants can now tolerate.

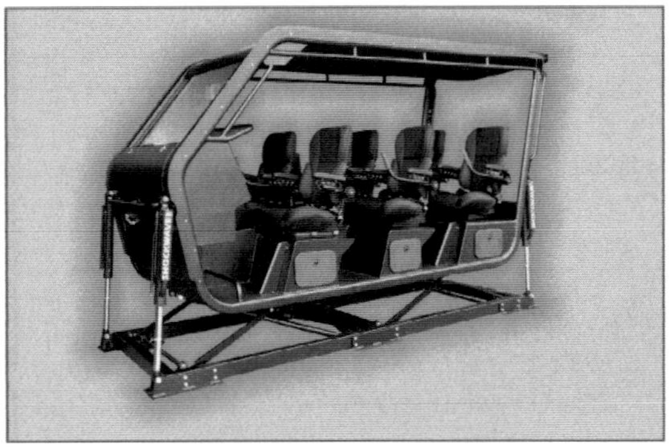

Figure 10: Integrated Crew Environment (ICE) Suspension system links controls and crew. Image Credit: Shockwave

20. IMPROVING SEAT CUSHION PERFORMANCE AND COMFORT

Foam materials have traditionally been used in marine seating to create padded cushions with the focus on comfort rather than protection from impact. On RIBs these tend to be foam pads ranging between 100mm (4") and 200mm (8"), with radiused top edges, that are covered in vinyl upholstery then fitted on top of seat boxes constructed of GRP / FRP or aluminium.

When the occupant sits on a foam seat pad it is immediately compressed, this compression increases during a transit and affects the cushioning properties of the foam and offers minimal protection from impact. Close cell foams are better than open cell foams for resisting water ingress, however over time all types of foam deteriorate losing their memory and shape recovery properties.

Trapped or pressurized air bag cushions usually require setting up to a specific pressure and adjustments due to air temperature.

Where vessels are unable to accommodate mechanical suspension seating due to weight / footprint / cost there is a requirement to improve the shock mitigating properties of the seat cushion. Utilizing a combination of polymer structures and geometry-based technologies SKYDEX Technologies have developed seat cushions to reduce the effects of impact and shock experienced during all modes of transport.

On planing and high speed vessels, including open RIBs and PWC (Personal Water Craft), laboratory and on water testing metrics demonstrate that improved protection for seat occupants can be achieved within the existing seat footprint and with no weight gain. Designed using highly durable materials SKYDEX relies on its polymer structures, not trapped or pressurized air, for cushioning and impact absorption. Requiring no set up or adjustments and performing in multi-axes, which improves shock mitigation and comfort in a wide range of sea conditions, the cushions are designed for marine use in harsh and demanding conditions. SKYDEX seat cushions are available as a retrofit cushion on all types of fixed seating or to further optimize performance of mechanical suspensions seats.

21. SEAT CUSHIONS AND IT BONES

In drop testing for mechanical suspension seats, varying masses are attached to the seat top and the seat is dropped from varying heights. From this, relevant metrics have been obtained to identify each seats shock mitigation characteristics in the laboratory or workshop before installation on boats for on water testing. There is now a requirement to develop a relevant standard for benchmarking the characteristics of existing foam cushions, and testing the next generation of cushion materials that have shock mitigating properties.

A major challenge with seat cushion testing is differentiating between the relatively large contact area of the human posterior and the relatively small impact area around the Ischial Tuberosity (IT) or sit / sitz bones. These are the protruding bones covered by flesh and muscle tissues of the gluteal muscles that contribute to the shape of the buttocks. The IT bones bear the weight of the body when sitting upright. Maintaining good posture is important on high speed craft. The spine needs to be aligned to reduce the risk of wave impact and repetitive shock causing disc or vertebrae damage.

22. BICYCLE SEATING AND MARINE SEATING

The racing cycle industry has utilised high tech materials to reduce weight, increase stiffness and raise speeds. As cycling grows in popularity both high performance professional and leisure riders are seeking improved saddle comfort with elements of the research relevant to marine seating. Although the surface area of a bicycle saddle is smaller, the relaxed upright cycling position (not racing position) is similar to the seating position on saddle shaped marine seats. Riders need correctly positioned handlebars to maintain good posture and marine seat occupants need correctly positioned hand holds for bracing and additional body support.

With the aim of reducing saddle discomfort, Italian saddle manufacturer Selle Royal commissioned a series of anatomical and ergonomic research projects with the German Sport University of Cologne. Researchers applied theory data and methods in an attempt to understand the fundamental design characteristics which optimise well-being of the rider and overall saddle performance. They have considered saddle shape and gender differences. A goal of the Ischial Variation Study is to determine and analyse the variety of IT bone distances, which may be relevant to marine seating design.

Researchers took various measurements and observations from 240 individuals (120 female and 120 male volunteers) to represent a cross-section of the population. The study included pressure mapping of saddles in use in different ride positions and recording the distance between ischial bones. The researchers found that male spacing varied between 7cm and 15cm with an average of 11.58cm. Female spacing in the sample measured varied between 9cm and 17cm, with an average of 13.4cm. Selle Royal concluded that, combined with the results from the pressure mapping gender study, individual ischial measurements had a bigger impact on comfort than gender.

23. SEAT CUSHION COMFORT VS PERFORMANCE

A further challenge with testing of high performance seat cushions is differentiating between the sensation of initial comfort and the objective of long term comfort, and for some designs impact reduction. With certain testing protocols a hard seat cushion may give the best metrics in individual drop tests, but be the least comfortable.

It is also important to ensure that seat cushions are not too soft, especially when transiting longer distances. Excess movement can exert pressure on the human posterior causing increased discomfort. Seat cushion designs allowing air movement and ventilation reduce the build up of heat on and in the cushion. This reduces sweating and the incidence of chaffing leading to skin rashes.

24. COMFORT IS THE ELIMINATION OF DISCOMFORT

Healthy skin tissue that is at the correct temperature, with good blood flow and no chaffing could be described as 'comfortable'.

A short steep sea, typically caused by wind against tide conditions, may not result in single large impacts but seat occupants may be subjected to continuous up and down movements on the seat every few seconds. This can cause reduced blow flow, increased temperature from pressure and friction which also causes skin chaffing and rash. In combination these could be described as 'uncomfortable'.

In addition to experiencing discomfort from up and down movements on the seat every few seconds, occupants may become more aware of IT bone pain from repeatedly landing on the seat surface. To reduce these effects seat occupants should be trained to stand every 10 to 15 minutes to stretch the back and legs muscles. Just as frequently moving hands around prevents numbness and pain, getting off the seat will also relieve constant pressure and improve blood flow.

Comfort is hard to measure and varies greatly from person to person due to anatomical differences, physical conditioning and pain tolerance. An approach to marine seating is that 'comfort is the elimination of discomfort'.

25. A HOLISTIC APPROACH TO HIGH SPEED BOAT OPERATIONS

Overall the fast boat sector now has a better understanding of the problems associated with high speed transits. Maritime sectors with high levels of exposure to open sea conditions include offshore energy, search & rescue, fisheries protection, navy and Special Forces. These organisations recognise that a shock mitigation strategy needs to be incorporated into design, training, planning and operations.

Technical shock mitigation solutions include innovative hull forms to improve ride quality, hull appendages to control angle of attack in waves, cushioned decks to stand on and suspension seating to sit on. Even with all this technology, important factors for individual mission success and long term capability are how an organisation takes care of passengers and the skill of the coxswain to drive according to the sea conditions.

A core component of fast boat training should be specialist knowledge to educate coxswains and crews to understand the forces that affect a planing craft, particularly when operating in waves.

For planing craft there are three distinct conditions to consider – displacement speed / getting on or off plane / planing speed. Where a craft follows the waters surface, for example when loitering in a stationary position or at slow speed in waves, the human response is unlikely to be injury but could be motion sickness. An issue with operating at 'hump-speed', where the boat is just below planing speed, is that the helmsman has poor visibility over the bow, which can reduce awareness of sea conditions. This in itself can give a hard and uncomfortable ride.

26. CONCLUSIONS

A holistic approach to shock mitigation solutions on planing craft can reduce the effects of shock and vibration on the human body and improve comfort for crew and passengers in all maritime sectors.

There are various means to reduce the effects of shock and vibration on occupants of fast craft. Although reducing the effects of RS and WBV by a few percent may seem minor, when a number of positive changes are made the cumulative difference can significantly reduce the incidence of injuries for boat crews and passengers. This can minimise downtime from injuries for the individual, extend the working life for groups of maritime professionals and improve operational efficiency for organisations.

If the risk of injury become too great for occupants it may be necessary to consider removing the man from the vessel. As USV (Unmanned Surface Craft) move from concept to in-service the maritime world is starting to realise that 'unmanned' may be the ultimate shock mitigation solution.

27. ACKNOWLEDGEMENTS

The author would like to express great appreciation: To KPM Marine, Scot Seats, SHOXS, Shockwave, SKYDEX Technologies for various background information and images. To the commercial, military and professional maritime sector in the UK, Europe, Canada, the US who have attended workshops and training courses then generously shared their experiences to improve the knowledge of shock mitigation on fast craft.

28. AUTHORS BIOGRAPHY

John Haynes holds the current position of Managing Director at Shock Mitigation Ltd. Maritime expertise includes 30 years training, consultancy and strategic product development focussed on the professional sub IMO / sub 24 metre sector. He is an Associate Fellow of The Nautical Institute, Associate Member of RINA, Commercial Yachtmaster Ocean and Advanced Powerboat Instructor. Shock Mitigation provides RS & WBV Awareness Training for the maritime sector: www.shockmitigation.com He is founder of the RIB & High Speed Craft Directory that brings together specialist boats and equipment for the professional fast craft sector worldwide: www.ribandhsc.com He has written specialist articles on future requirements and new technology for over 20 international publications. At Seawork 2016 he was voted Maritime Professional of the Year.

HOME IMPROVEMENT – STICKY BACK PLASTIC FOR WIND FARMS? A SAFETY RELATED APPROACH TO WIND TOWER SOV APPROACH INSTRUMENTATION

A Stead, Guidance Marine, UK
H Busshoff, Bernhard Schulte Ship Management, Germany
Christoph R Grams, 2nd Officer, M/V "WINDEA La Cour", Germany

SUMMARY

This paper summarizes some test carried out by the Windea La Cour SOV in early 2017 with a view to optimising Position Reference Sensor (PRS) target design and location for wind farms. Finally we discuss how the new target will be able to mitigate all false reflections when used with the CyScan AS sensor.

NOMENCLATURE

PRS	Position reference sensor(s)
SOV	Service Operations Vessel
WLC	Windea La Cour (an SOV operating in the Gemini field)

1. INTRODUCTION

Until new target-less Position Reference Sensors (PRS) such as the RangeGuard Monopole[1] start gaining acceptance, targeted systems, particularly laser PRS, will continue to dominate where SOVs are positioning off wind turbines when deploying gangways.

The number of landing stages on each wind turbine coupled with the number of turbines in the field necessitates significant numbers of targets to be installed across the wind farm. Some operators have chosen low CAPEX solutions that will require regular maintenance or upgrades. Others have chosen higher CAPEX solutions in the expectation of a lower lifetime cost[2]. These choices have had to be made with imperfect and incomplete safety data leading the industry to adopt a "best guess" approach to installing targets.

This paper looks at tests carried out by the Windea La Cour crew on alternative targets with a view to informing the debate on best practice for Wind Farm PRS target installations.

2. TECHNICAL BACKGROUND

Laser PRS work by horizontally sweeping an asset with electromagnetic energy fanned out along the vertical, the swept area being sufficient to find any retroreflective targets. The retroreflective nature of the target sends the signal back to the sensor and the measured range (by time of flight) and bearing (by rotary encoder) are used to determine relative position. Two targets fixed to the installation, with sufficient angular discrimination will additionally give the relative heading. All PRS manufacturers, in a bid to maximise range, remain eye safe, and keep costs down, use 905nm light in the near infra-red part of the spectrum as their primary energy source.

When selecting suitable targets wind farm designers and operators must choose an engineering compromise to optimise safety whilst controlling costs. The following lists some of those considerations and identifies those that were tested by the WLC.

2.1 TARGET MATERIAL CONSIDERATIONS

Laser PRS targets generally fall into two types: retro reflective tape (sometimes known as "tubes") and glass prism based targets. Both have distinctive characteristics that affect both safety and price.

2.1 (a) Price

In general tape based targets are about 1/3 the price of glass prism based targets. Depending on the size and quality of the materials used a tape based target can vary from €200 - €800 whereas a prim based target can vary from €600 - €2500.

2.1 (b) Degradation

All targets will degrade in harsh offshore environments. Extremes of temperature combined with the constant present of moisture, corrosive salt and UV degradation present significant challenges to target design. Local environmental challenges, such as statically charged snow or sand can also lead to obscuration and/or surface degradation.

This trial has not tested targets over time relying instead on quality targets from reputable manufacturers. For a given quality standard, tape targets are generally expected to have about 1/3 the life of a glass prism target.

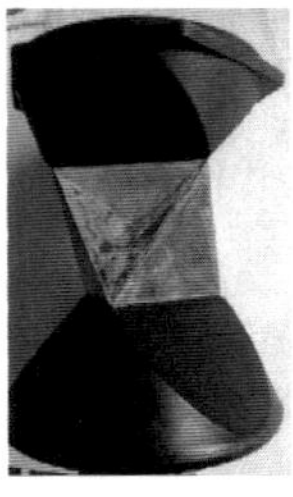

Figure 1: Example of a lower cost glass prism after three months at sea. The reflective backing can be seen peeling from the back of the glass. Targets designed for land based surveying can be expected to degrade in a similar manner.

2.1 (c) False Targets

It is important to note that *any* retro reflective material will act as a potential target. This can be a particular issue on small assets such as wind towers where such "false" targets (reflective jackets, shiny toolboxes, backs of lights etc.) must, by necessity, be co-located with the real targets.

A sensors ability to distinguish between these "false" targets and the real ones is a key safety consideration and depends in large part on the properties (and consequent costs) of the selected target. This is considered as part of the trial.

It is worth noting that PPE material which has previously been compliant to ISO class 20471-2 is progressively moving to the "brighter" and more reflective ISO20471-3 standard. This will increase the likelihood of false reflections. Guidance Marine recommends the use of prism reflectors as the only safe solution suitable to work with laser PRS in offshore wind[3].

The WLC trial compares tape targets to prism targets.

2.2 TARGET LOCATION CONSIDERATIONS

The targets can be located anywhere provides they are: visible, easy to clean, on a fixed baseline to the landing stage (can also include being on another wind tower).

The trial sought to look at various mounting options.

2.2 (a) Line-of-Sight

It is important that laser PRS have clear line-of-sight to the targets at all times. This can be a problem where gangway deployment will inevitably obscure areas proximate to the landing stage. The circular nature of wind farm platforms can also make it difficult to see a target from every angle of approach.

Figure 2: View from traditional sensor location. Target can be clearly seen if approaching in right direction.

In general the sensors on the SOVs are mounted, for historical reasons - achieving line-of-sight to high mounted oil installation targets - at bridge level. This has the effect of making them more or less level with the landing wind tower stage and the vessel gangway on an SOV (figure 2). Advantages of this includes putting the sensor within clear line of sight of the handrail which provides a simple target mount which is easy to access for cleaning. A disadvantage is that the gangway is working at that level and runs the risk of obscuring the target. Sensors working lower on the vessel can get a clear all round view of the wind tower (figure 3).

Figure 3: View from gunwale.

The trial was unable to do much work exploring optimal locations but some conclusions are drawn.

2.2 (b) Range

Whereas the glass prisms retro-reflect almost all the incident light with a high level of directionality, tape targets will reflect much less of the signal (much will be absorbed by the substrate) and that which is reflected will have low directionality. Being cheaper, it is possible to create a much larger target and thus reflect more of the fanned beam but the lack of directionality will cause that signal to attenuate faster than the prism targets impairing the range.

The trial offers some data on signal attenuation over range.

2.2 (c) Angular Coverage

Along with range and line-of-sight the angular coverage will affect the usability of a target and the speed and safety with which an SOV can reach position and deploy the gangway.

Tube targets have the advantage of offering full 360 degree coverage whereas and individual prism will cover about 60 degrees. The prism target used in the trial consists of a cluster with four overlapping prisms across the horizontal offering 190 degrees of coverage (figure 4).

The trial seeks to identify the minimum angular coverage needed.

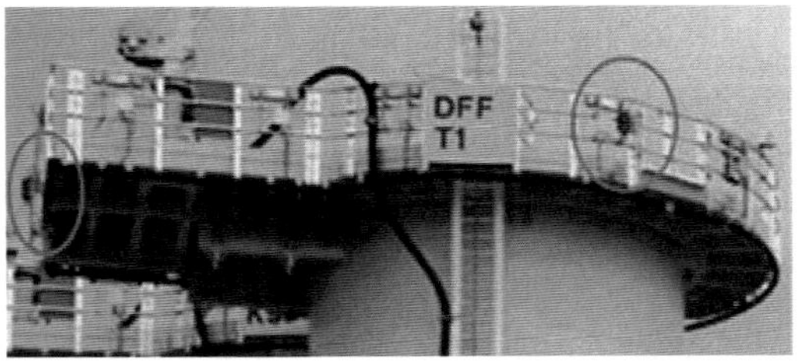

Figure 4: 8 way 190 degree rugged prisms on the Dudgeon TPs. A third unit is mounted on the other side, providing full coverage.

3. THE WLC TRIAL

The tests took place over two separate days 30/1/2017 and the 01/02/2017 while the WLC was working in the Gemini wind farm. By necessity the tests had to be implemented around the working schedule and with limited and imperfect control of the test environment. Nonetheless our thanks go to Christoph Grams and the crew of the WLC. The results are presented, as found, to add value to the debate and inform further tests.

3.1 OBJECTIVES

Accepting that Prism based targets provide a more secure target than tape based targets the objective of the trial was to see how much cost could be removed from a prism target without impacting performance within a wind farm environment.

The trial involved taking Guidance Marine's 8 prism Rugged Prism target and eliminating costs (by removing prisms) and testing the operational effect of each downgrade.

3.2 TESTS

3.2 (a) Test 1: Tubes Versus Prisms

Method:
1. Hang an 8-prism cluster on the handrail in the best location (figure 5)
2. Assess the performance of the prism during departure and approach.
3. Repeat on several turbine approach/departures

Results:

"The first trial with the Prism was very successful. The CyScan locked on to the Prism very well ignoring the installed tubes. Once the Prism was covered by the gangway, the CyScan jumped to one of the tubes.

CyScan didn't jump the target when locked on to the Prism, when people with reflective clothing were walking by, as has happened before with the tubes. The signal from the prism was still good at 100 m distance. Brightness 90%-very good compared to tube 15%.

In any case we will need 3 Prisms on every Turbine, due to the Gangway operation if mounted at the balcony level"

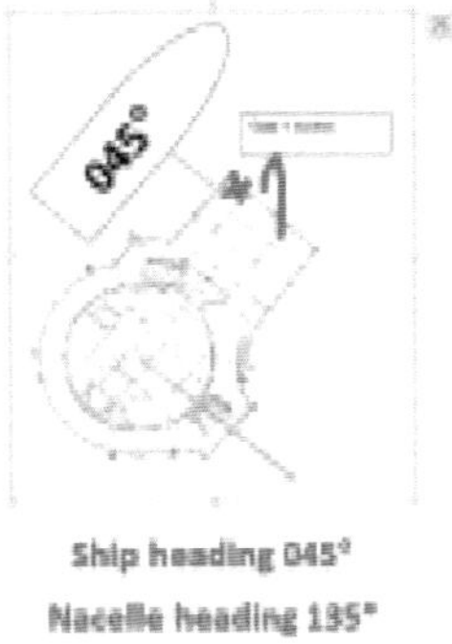

Figure 5: Prism location, test 1.

Other comments:

"CyScan is not jumping from the target to another target at all, even if the minimum brightness is set to ZERO. The reception of the target starts very early at 150m depends also a bit of the angle to the prism."

3.2 (b) Test 2: Test Different Locations.

Method:
1. Hang the 8-prism cluster on a turbine in an alternative position
2. Assess the performance of the prism during departure and approach.
3. Repeat on several turbine locations during approach/departures

Results:

The WLC were unable to complete this test. The only results were from the first location used on each test. Below is a general comment from the WLC:

"If the Prism is mounted at the balcony you need 3 prism to cover all gates for approach because it gets covered by the gangway.

If they will be mounted on a higher level above the gangway 2 prisms might be sufficient."

Note that the same could be said of mounting on a lower level below the landing stage

3.2. (c) Test 3: Cover Individual Prisms to Establish the Minimum Angular Coverage Necessary.

Figure 6: Prisms circled in red remain uncovered. Vertical configuration.

Method:

1. Use the lens caps to cover 6 of the prisms, except for the 2 circled in red (figure 6)
2. Hang the prism on a turbine in the best position, determined by tests 1 and 2 (figure 7)
3. Assess the performance of the prism during departure and approach

Figure 7: Prism location, test 3.

Results

First attempt:

Test carried out some time during previous week.

"The prism target was placed at the corner of the Balcony to the right hand side of Gate 2. During approach to gate one the two uncovered Lenses were not visible. The Prism target was not brighter than the tube.
If approaching the turbine from the side, it is better to have lenses all around the target. The Vertical Configuration is not ideal for our operation."

Second attempt:

Test carried out 31.01.17 0900LT

"The problem was, at short distance around 16m it was losing/jumping on the tube. As soon as we moved further out the target became stable.*

At a distance of 20m the target became stable. We moved away parallel to the landing stage up to 250m"

The brightness of the target versus distance is shown in table 1.

Table 1: Brightness vs distance, vertical configuration

Distance [m]	Brightness
40	80%
55	75%
100	70%
150	60%
250	40% Slowly loosing target.

"For the full passage outside it kept track and did not jump to another target. It was a good result.
The signal was generally much better than the tube.

At a distance about 150 m we moved ahead and astern to check where Cyscan loses the target. Cyscan lost track at a side bearing of more than 135degrees when we moved ahead of the Turbine.

This configuration is not really suitable for our kind of operation, due to different weather approaches. We are also approaching from ahead and astern towards the turbine."

* The logs shows that this was due to saturation of the sensor receiver. At such short ranges the sensor was unable to discriminate between the tube or the prism. A modification is under development.

3.2 (d) Test 4: Cover Up Individual Prisms To Establish The Minimum Range Necessary For A Safe Approach.

Method:

1. Use the lens caps to cover 6 of the prisms, except for the 2 circled in Yellow (figure 8)
2. Hang the prisms on a turbine in the best position, determined by tests 1 and 2 (figure 9)
3. Assess the performance of the prism during departure and approach

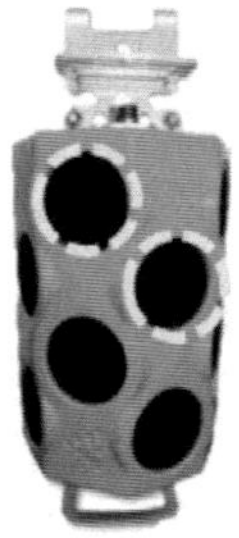

Figure 8: Prisms circled in yellow remain uncovered. Diagonal configuration.

Results:

Test Date: 01.02.2017

"With test No 4 and the diagonal configuration CyScan never lost track of the prism. It did not jump to the tube at all. We moved away parallel to the landing stage up to 250m.

The brightness of the target versus distance is shown in table 1.

Table 2: Brightness vs distance, diagonal configuration

Distance [m]	Brightness
40	70%
55	65%
100	65%
150	60%
250	40% Slowly loosing target.

"This configuration looks more suitable for us than Test 3.

Figure 9: Prism location, test 4.

Along ship parallel movement did not interfere the tracking at all. We moved back and forward to test if CyScan is losing track when the side bearing reaches a certain value. He kept on tracking."

4. CONCLUSIONS

The test seemed to corroborate the expectation that prisms would outperform tubes:

"CyScan didn't jump the target when locked on to the Prism, when people with reflective clothing were walking by, as has happened before with the tubes."

The range from a single prism layer appears adequate for the purpose of deploying a gangway. In fact, a single prims may be better for very close proximity work (minimises the chance of saturation).

The need for good angular coverage was clear:

"CyScan lost track at a side bearing of more than 135degrees when we moved ahead of the Turbine. This configuration is not really suitable."

Figure 10: 4-prism cluster that covers 180 degrees.

It would appear that a 4 prism cluster covering the full 180 degrees but with just a single prism on the vertical would give the optimum price versus performance (figure 10). It would halve the cost of prism targets whilst only marginally impacting range.

Proposed OEM low price (<800 €) rugged prism suitable for wind farms

5. NEXT STEPS

A full trial with one tower instrumented with tubes, one with full eight way prism and one with the proposed new four way prism is proposed.

6 DISCUSSION – A DPO PERSEPCTIVE

From a DPO perspective, there are 2 major risks when using reflective targets with laser PRS.

- Target swaps or target jumps between legitimate reflector targets
- False reflections from other reflective objects in the field of view

Both of these occurrences could result in a sudden movement of the vessel as a new distance is suddenly inputted into the DP system. If either occurred at close proximity to the wind turbine, or when the walk to work gangway is deployed, the consequences could be devastating.

Although the use of prism targets dramatically reduces the risk of these events happening, due to the high brightness of the prism compared to other reflective objects in the vicinity, it does not eliminate the potential problems.

CyScan AS is a new patented technology from Guidance marine that will be launched in 2017 and can completely eliminate these risks[4].

The CyScan AS sensor has two lasers at slightly different wavelengths (rather than one laser) and the prism targets have light filters which absorbs one of these wavelengths. When a CyScan AS sensor is used in combination with an AS Prism target, only one wavelength is reflected from the prism and detected by the CyScan AS. The CyScan AS knows that this target can only be an AS prism target[5]. The prism has an "Absolute Signature". All other reflective targets will reflect both lasers. It should be also noted that no objects in the natural world will behave like AS Prism.

By combining AS technology with what we have learnt from the prism tests described in this paper, we can propose a cost effective prism cluster that provides additional safety. It is a solution that is tailored to the specific needs of offshore wind farm operations.

7. REFERENCES

1. SY Heriot, T Coggins, M Mudrinic & A Stead, H Busshoff, O M Husøy, A New Era in Local Position Referencing, Design & Construction of Wind Farm Support Vessels, 30-31 March 2016, London, UK

2. Guidance Marine, 201 Rugged Prisms to be installed on the Dudgeon Offshore Wind Farm project, http://www.guidance.eu.com/UI/Content/Content.aspx?ID=551

3. D Sanderson, PRS Reflector Target Recommendation for the Offshore Wind Industry, http://www.guidance.eu.com/assets/_managed/cms/files/PRSReflectorTargetRecommendationfortheOffshoreWindIndustry.pdf

4. R Miles, Laser Target Ambiguity, Dynamic Positioning Conference, 11-12 October 2016, Houston. http://www.guidance.eu.com/assets/_managed/cms/files/LaserTargetAmbiguity.pdf

5. Guidance Marine, Absolute Signature - AS Technology is Coming, March 2016 http://www.guidance.eu.com/assets/_managed/cms/files/94-0495-4AbsoluteSignature_16.pdf

8. AUTHORS BIOGRAPHY

Andrew Stead is Head of Business Development for Guidance Marine. He began his career at Campbell Scientific in the in the sensors industry as Head of Sales supplying data acquisition systems. In 2008 he was appointed Managing Director of Muir Matheson supplying MetOcean and Helideck monitoring systems to the oil and gas industry. In 2012 he joined Guidance Marine as Head of Sales.

Hendrik Busshoff holds the current position of Offshore Marine Superintendent at Bernhard Schulte Ship Management. As a trained Master mariner he is responsible for offshore marine operations and DP. His previous experience includes SDPO on hyperbaric dive support vessel in the North Sea with Technip and sea time as officer and Marine Superintendent with the merchant navy.

Christoph R Grams is the 2nd Officer on the SOV "WINDEA La Cour". A certified Dynamic Positioning Officer and Helicopter Landing Officer he has previously worked in offshore in seismic survey.

NOMOGRAM OPERABILITY CURVES FOR WIND TURBINE INSTALLATION VESSELS – BENEFITS AND PITFALLS

Z Rahman and **A M Raithatha**, Noble Denton marine services, DNV GL

SUMMARY

Owners and Operators of Wind Turbine Installation Vessels (WTIVs) are attempting to extend their understanding of the operating limits of their vessels beyond the limited number of 'design' case combinations to assist in the tendering process and maximise operational efficiencies to drive time and cost-savings during installation.

A useful tool to assist with this is the development of operability curves (or "nomograms") considering a range of conditional load combinations to supplement the vessel design conditions, which are typically based on a single set of parameters. With careful consideration of parameters, nomograms have potential to be a valuable tool for both tendering processes and supporting location approval for units to operate on-site.

DNV GL worked with WTIV owners and operators to develop operability envelopes using bespoke iterative scripts and WTIV models using their in-house "JUSTAS" analytical software. The envelopes have been summarised in spreadsheet format to allow WTIV owners and operators to interactively determine their unit's operability for user-specified input conditions and therefore help optimise both the installation sequence and loading conditions for operations.

This paper presents brief background to the topic and examples of nomograms in the context of jack-ups (both Oil & Gas and WTIVs) with discussion of not only the perceived advantages, but also a note of caution on the potential pitfalls and where they might not be suitable for use.

NOMENCLATURE

LPA	Load Penetration Analysis
MWS	Marine Warranty Surveyor
OWF	Offshore Wind Farm
SSA	Site-Specific Assessment
WTIV	Wind Turbine Installation Vessel
WTG	Wind Turbine Generator

1. INTRODUCTION

The term "Nomogram" originates from the Greek words nomos (meaning "law") and grammē (meaning "line"). From the Oxford English dictionary, a nomogram is a diagram presenting the relations between three or more variables using a number of scales, arranged such that the value of one variable can be found by drawing a straight line intersecting the other scales at the appropriate values. As an example Figure 1 from AISC steel manual [4] shows a nomogram for deriving the effective buckling length factor, 'k', of columns in moment frames where 'k' is determined from the point of intersection of a line between known values of G_A and G_B. An example is depicted as a dashed line in the figure, for G_A=1 and G_B=10, giving k = 1.9.

2. USE IN CONTEXT OF JACK-UPS

When a jack-up or WTIV operator seeks to install their unit at a particular location, the location approval of the vessel is subject to a review being completed by a marine warranty surveyor (MWS) [1]. One aspect of this review includes a storm survivability assessment in accordance with criteria such as ISO 19905-01 [2] or SNAME T&RB 5-5A [3].

Key factors that will be checked in this assessment include the intended loading configuration (weight, cargo, windage and airgap), waterdepth and the foundation and environmental conditions. These are used to judge whether a proposed installation site is suitable for operations or not. These checks may be undertaken by (c.f. [1]);

- Reviewing of site and loading conditions against designer-rated parameters, which can be in the form of Nomograms
- Performing a site-specific analysis (SSA) to assess unit capability specific to the location of interest
- Using a previous SSA as the basis for a comparative review

The use of the term "Nomogram" in the context of jack-ups, historically relating to the Offshore Oil & Gas industry, refers to an operability graph typically plotting the allowable leg-load for the jack-up against allowable wave-height with separate curves for a third variable.

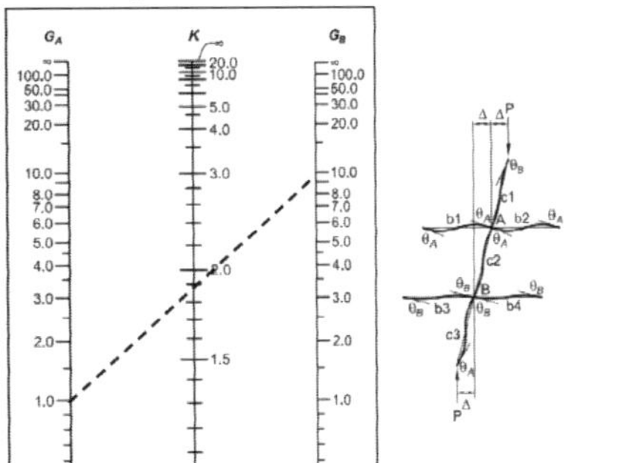

Figure 1 – Example Nomogram for deriving effective buckling length of columns in moment frames

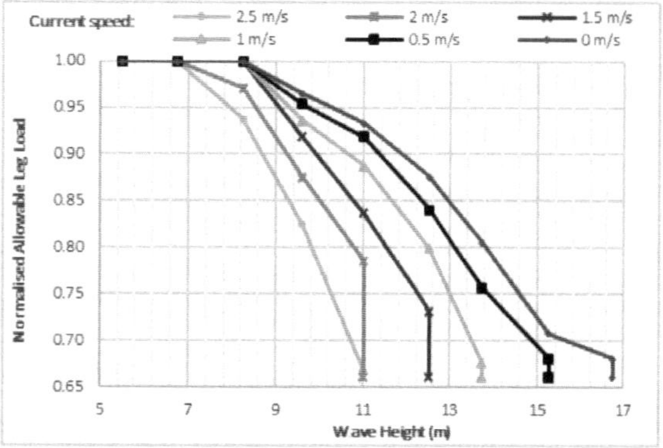

Figure 2 – Example of a nomogram for a jack-up

"Allowable leg-load" represents the maximum stillwater weight condition per leg that the jack-up may adopt to survive a given storm condition. An example is provided in Figure 2, where each separate curve applies for a different surface current speed. The choice of the three variables presented on a Nomogram can, just as easily, be allowable wave-height against allowable water-depth with separate curves plotted for e.g., wind-speed.

Due to the deviation in use of the term "Nomogram" from its historical origins (i.e. from the form illustrated in Figure 1 to that shown in Figure 2) the latter may also be referred to as a "Nomogram Operability Curve" for extra clarity.

The use of jack-ups in the Offshore Wind Industry has developed relatively recently with the industry itself still in its early stages relative to the Oil & Gas industry (UK Government and Crown Estate granted permissions for the first seventeen 'Round One' windfarms as recently as 2003 [1]). The format of Nomogram Operability Curves in the context of Wind Farm jack-ups has therefore replicated their use for jack-ups in the more mature Offshore Oil & Gas sector.

3. BENEFITS OF USE

3.1 Simplicity of interpretation

One aspect of nomograms that makes their use appealing is the ease with which they may be interpreted by, for instance, those without a specialist engineering background, extensive experience in jack-up operations and/or access to bespoke software.

Despite the best intentions of engineers, it is simpler for the non-specialist, for instance the marketing department of a vessel or the procurement department of an operator, to read the operability of a jack-up graphically rather than from detailed Site-Specific Assessment (SSA) reports (which are discussed in Sections 4 and 5).

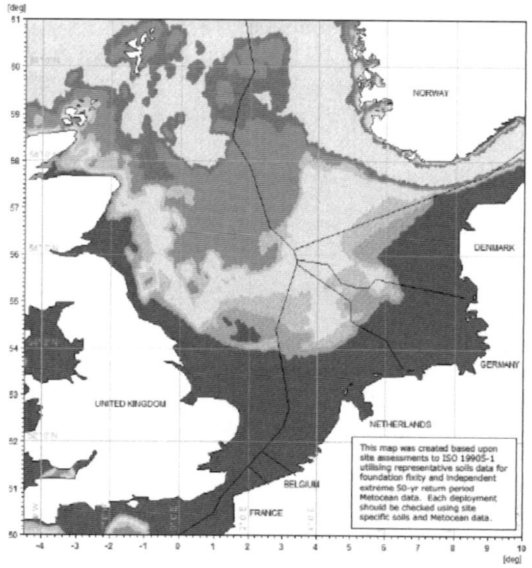

Figure 3 – Example of a North Sea Operability Map for an Oil & Gas jack-up derived from Nomogram Operability curves

3.2 Pictorial representation

A useful by-product of nomogram operability curves is that they can be combined with environmental contour maps for a region of interest to create an operability map for the jack-up (see Figure 3). This type of operability map demonstrates clearly the benefit of ease of interpretation, with Green areas representing the greatest all-year operability and Red representing the lowest summer seasonal operability.

The nomograms can therefore present operability limits for a range of operating conditions on a single page and this can be an advantage over an SSA report. Due to the large area depicted and hence low spatial resolution, this type of map is best suited for use in marketing and "pre-screening" activities in the tendering phase, where it can be sufficient to have a general overview of the acceptability of a location.

3.3 Interactivity via Spreadsheet/Computer program

Nomograms for a WTIV can be stored as a database in a spreadsheet or computer program designed such that, for a user-defined vessel operating point such as water-depth, storm wave-height or wind-speed, the applicable nomogram is presented to the user, with the user's operating point over-plotted on the nomogram to give a simple graphical result of whether the WTIV is a pass/ fail for that particular case. An example is shown in Figure 4, where the range of operability is bounded by the solid black line and solid blue line. The operating point in this example is denoted by the red cross and this example shows that it "passes" the storm survival criteria as it is inside the "acceptable" zone.

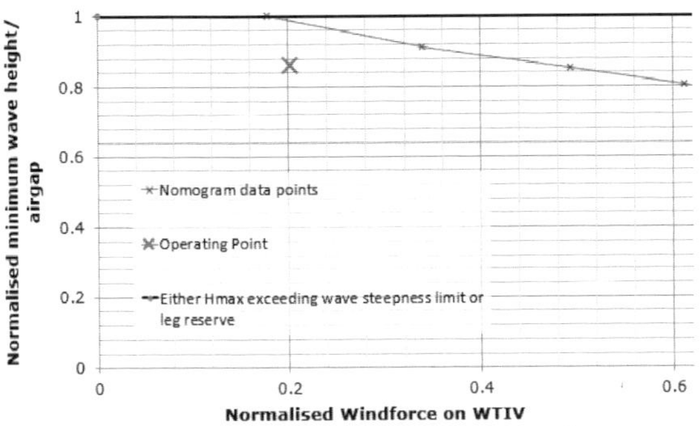

Figure 4 – Example WTIV Nomogram screenshot

The main advantage of such software is that it provides an ability for a non-specialist end-user to input data, interpret the results and quickly check a range of cases.

3.4 Optimisation of operations

It has been discussed that a nomogram allows a jack-up owner or operator to make an early, but well-informed, decision about whether an installation site is suitable for their vessel or not (e.g. using maps like Figure 3).

Further along the project cycle and specifically relating to OWFs, such operability curves have potential for the jack-up owner or operator to optimise their installation sequence. For example, for a WTIV operating on an OWF comprising 20 WTG Installation sites, it is expected that environmental extremes will be uniform for the entire OWF. Ignoring foundation conditions, which are treated with conservative restraint conditions in a nomogram, the only site variable for each individual WTG site in an OWF is the water-depth.

A nomogram presenting allowable wave height against allowable water-depth can be used to rank locations by severity of jack-up utilisation. If, for example, it is shown that a location with deeper water-depth has a reduced allowable wave-height then a WTIV owner or operator may choose to schedule this location last in an installation sequence to benefit from reduced cargo weight and wind load. Similarly, if operations at an OWF are expected to last several months, the sequence can be optimised such that the most severe locations in terms of storm survivability are visited during the most favourable months.

4. PITFALLS OF USE

4.1 Generic modelling requirement

Typical nomograms consider three varying parameters; e.g. wave height, water depth and wind-speed, with other parameters e.g. current speed, foundation conditions, airgap, cargo weight configuration and windage fixed. To ensure the nomogram is representative for a large enough range of those fixed parameters, the fixed parameters may need to be set conservatively.

Several conservative assumptions could be compounded together, such as a maximum anticipated current speed together with a pinned foundation model and heaviest deck loading condition. Consequently the operability of a WTIV, as presented by a nomogram, is expected to be more conservative than could be demonstrated by assessing individual site conditions on a per-case basis, and in some instances the nomogram may be deemed overly conservative.

4.2 Inability to model in further detail

To generate a nomogram operability curve requires a large amount of background analysis and may involve iterations of hundreds or even thousands of cases, to generate sufficient data points for a reliable curve. The computing effort needed has historically been a limitation which made suitably detailed nomograms intractable. This has been alleviated by increased accessibility to computing power and the continual improvement of bespoke and efficient software for assessing jack-ups in their elevated condition, such as DNV GL's JUSTAS program.

However, once nomograms have been generated and published, it may not be a simple matter to undertake detailed further investigations to eliminate conservative modelling assumptions and derive greater operability. In contrast, many options can be assessed in a short time by a specialist undertaking an assessment for a specific location.

Such assessments involve developing a model of the site-specific foundation conditions, the global structure of the jack-up and its configuration including cargo weight, windage, rig heading and airgap. The model is subjected to environmental loading and a range of strength checks or 'utilisations' are calculated, for key parameters such as leg structural strength or foundation bearing capacity, to judge whether a proposed installation site is suitable for operations. Details of the analysis are documented in an SSA report. Some of the tools that can be employed in such an SSA are:

- **Directional assessment** – Nomograms are traditionally derived using omni-directional extremes which are applicable across multiple locations, with the maximum storm condition applied uniformly from all headings around the compass. In an SSA the directionality of the site-specific metocean extremes can be considered, with the rig-heading aligned favourably such that loading for the most extreme storm heading is as small as possible.

- **Site-specific foundation modelling** – For nomograms to be applicable for a range of locations it is necessary to apply a simple foundation model in them. Generally a pinned foundation, or notional level of foundation fixity is assumed. The resulting foundation model may be conservative for many locations. The foundation capacities and fixities that are calculated from site-specific soil data in an SSA are typically less onerous and more representative of a particular location of interest.

- **Additional settlement under storm loading** – Depending on the foundation condition if footing loads exceed the load applied during installation there is potential for additional penetration to be realised causing a tilt of the hull. Overutilisation of foundation bearing capacity checks will therefore be subject to review of the load penetration response of the soil for a particular location to determine if the potential additional settlements may still be within acceptable limits. Such consideration of additional settlements are an example of removing unnecessary conservatisms that can only be considered in a site specific basis.

- **More detailed structural modelling** – From a nomogram it is often not clear or known which parameter is governing (such as foundations, holding system or leg structure). In an SSA, utilisations for various measures are reported individually, and if it is found that the unit is limited by, for instance, a structural over-utilisation then it is possible to undertake further detailed analysis to remove local conservatisms and try to find additional operability. In the case of a nomogram, such a modelling change can be made but it then requires re-assessing the entire matrix of hundreds or thousands of cases to replot the nomogram, which removes the objective of producing the nomograms in terms of simplicity and usability for a range of parties.

4.3 Optimised preload/predrive

A key benefit of SSAs is the ability to reduce the target predrive reaction for the vessel when a WTIV is shown to comfortably pass storm survival assessment criteria (e.g. ISO [2] or SNAME [3]).

Preload (addition of additional weight by filling ballast tanks) or predrive (jacking a four-leg unit to place the elevated load on a single diagonal pair of legs) is the vertical proof load that a WTIV applies to the seabed when it is installed at a location. The purpose of this is to proof test the soil to a certain vertical load to ensure that additional foundation settlement will not occur during the extreme storm. After preloading or predriving is completed the leg load is returned to the nominal gravity value by dumping preload ballast and/or sharing elevated weight equally among the four legs of a WTIV. This condition is called the stillwater footing reaction on each leg of the WTIV.

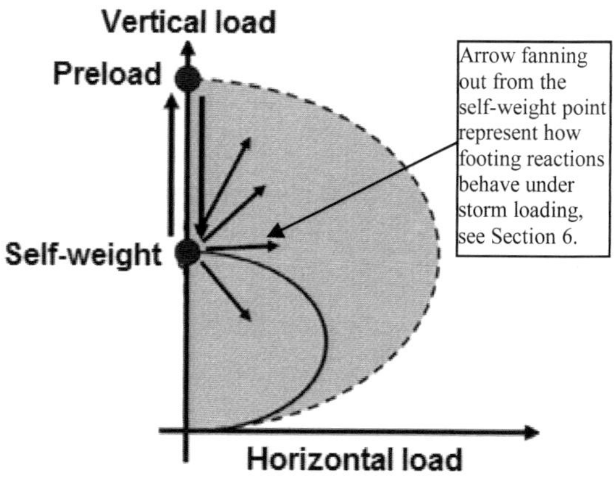

Figure 5 – Example of proof-tested Vertical-Horizontal foundation bearing capacity envelope created by applying predrive/ preload to soil

The principle is then that the additional vertical and horizontal load applied to a seabed when the WTIV is exposed to a storm, does not exceed the margin between stillwater footing reaction and the proof-tested predrive reaction (or preload). This is shown graphically in Figure 5, where the shaded region depicts the allowable vertical-horizontal footing load capacity envelope.

The installation or predriving process for a WTIV is one that can take several hours. Any reduction in this target pre-drive, derived from an SSA may save installation time significantly and therefore costs to the WTIV Operator in daily costs and/or the OWF Operator day-rates. There could also be a benefit of reduced wear on the jacking machinery by applying lower predrive loads.

5. NOMOGRAMS vs SSAs

5.1 Use in Location Approval for Marine Warranty

Given the advantages of using nomograms in the context of Location Approval for jack-ups described in Section 3, it is worth mentioning that there was a drive in the Offshore Oil & Gas industry in the 1980's to move toward using nomograms for Location Approval instead of SSAs. This would have been appealing especially for installation sites where the jack-up is expected to pass storm survival criteria comfortably, and the conclusions of a Nomogram and an SSA could be the same.

The early jack-ups in the Oil and Gas industry were designed using methods that could only consider a limited number of variables, making them amenable to nomograms, which as previously discussed, can represent operability against only three variables at any one time. As the industry developed, analytical tools and operational know-how evolved to allow jack-up structural designs to be refined against a vast array of load cases and design variables. There was also a push to find the boundaries of operability for Oil & Gas jack-ups, requiring installation at locations which were proportionately more severe in terms of water-depth and environmental extremes. To meet this requirement, the benefits of using SSAs to extract greater operability meant that their use became prevalent and they displaced the use of nomograms, which became used more for "pre-screening" and marketing activities.

A further contributor to this trend was the reduction in costs for performing SSA's through standardisation, increased industry knowledge and the development of increasingly powerful software to undertake the necessary calculations more efficiently.

5.2. Cost-comparison

The lead time for undertaking SSAs was aligned with operations for typical drilling jack-up rigs, which would typically attend less than 10 locations per year. Interest in nomograms has been revived by the windfarm industry, where WTIVs can jack up at over a hundred locations in a single year.

Under this context the economies of scale that could be gained from investing in the development of and using nomograms, against traditional SSAs is very attractive. Hence, it is perceived that by developing nomograms for Location Approval there will necessarily be a cost saving by removing the ongoing need to undertake SSA's for every location. Whilst this can certainly be the case, a cost-benefit analysis should be performed that considers the following points:

- Due to the large number of assessment cases (hundreds or even thousands) required to create a Nomogram Operability Curve there is an "up-front" cost to their generation.
- Due to their necessity to contain conservative modelling assumptions, for the more severe installation sites, which are expected to become more common as the industry develops by moving to deeper waters c.f. [1], it may become more often the case that the nomogram gives a negative answer for operability where a detailed SSA could give a positive answer.
- Nomograms do not bypass the need to complete some aspects of a site-specific analysis. A spudcan load penetration analysis (LPA) is still required for each jacking location in an OWF to confirm a WTIV's expected penetration depth under applied predrive and to identify potential installation hazards such as punch-through (or rapid penetration) risks.

Consider a hypothetical example. A windfarm containing ~50 WTG installation sites requires assessment of soil profiles for all 50 sites, and then LPA's for a reduced number of soil categories (we will assume six categories for the entire OWF). SSA's will only need to be completed for the reduced number of soil categories to cover all the locations in the OWF. Given that additional foundation studies are often required to mitigate installation hazards, the cost-contribution of SSAs to a typical OWF Location Approval project is usually less than 50% and understood to be often be closer to 10-20% of the overall cost.

This saving may or may not be significant depending on the specific context; it should be weighed-off against possible efficiencies that could be gained during operations from optimised installation sequence and reducing the magnitude of required predrive.

6. CASE STUDY

DNV GL's Noble Denton marine services section have been engaged by various operators to develop nomograms for their vessels that, where found applicable, could be used by a MWS as a basis for providing location approvals. This case study is an amalgamation of observations from undertaking these studies, with details normalised such that they do not relate directly to any specific project.

Early nomogram studies for WTIV's typically took place using the same approach as traditionally undertaken for the Oil & Gas sector, with conservative penetration depth, foundation capacity, pinned footing restraints, a heavy deck loading condition and the largest anticipated surface current speed. Assessments were completed against SNAME criteria [3] and for a range of anticipated wave heights, water depths, and wind speeds.

Python scripts were developed to generate separate input files for DNV GL's in-house JUSTAS model of the WTIV for each combination of wave height, water depth and wind speed, resulting in thousands of analysis cases. All the input files were run in sequence overnight using the batch function in JUSTAS and the resulting structural and foundation loads reported in output files were then also post-processed via Python scripts, to find the combinations of wave height, water depth and wind speed that resulted in the WTIV reaching its limit of acceptability (for structural and foundation loads) in accordance with SNAME criteria [3].

Example limits of acceptability include the leg reserve, leg structure and holding system strength, preload footing reaction, foundation sliding and foundation bearing capacities reaching a utilisation of 1.0.

Of these the foundation bearing capacity is discussed in further detail as, whilst it was included in DNV GL's recent studies, the authors have observed that historically it has often been omitted from nomogram studies. This omission is significant as it is often the case that the foundations reached their limiting utilisation before the leg and holding system structural utilisations reached unacceptable levels.

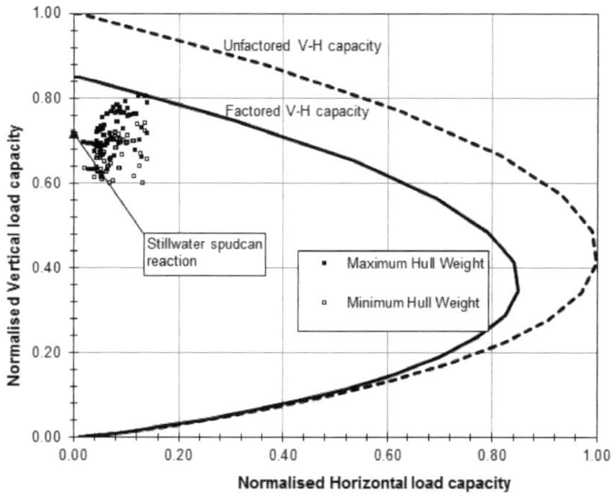

Figure 6 – Example of Foundation bearing capacity utilisation = 1.0 for a SNAME assessment

An example of a foundation bearing capacity check for one limiting combination of wave height, water depth and wind speed is provided in Figure 6. The capacity envelope represents a weak sand foundation, selected as it is also conservative for clay soils and hence representative for the widest range of locations. The solid black line represents the factored vertical-horizontal load bearing capacity envelope of the foundation derived using the SNAME equations. The solid black squares show the maximum storm footing reactions derived from the storm loading analysis for a range of storm headings for the maximum elevated weight condition of the vessel. Another set of loads are denoted by white squares, which represent a lower elevated weight condition of the vessel.

The storm footing reactions fan outwards from the stillwater footing reaction (i.e. the vertical load on the jack-up footings under static load alone and no horizontal reaction) denoted by a black triangle on the vertical axis. The maximum storm footing reactions consist of horizontal and vertical force. The storm footing reactions shown in the graph arise because all footings of the WTIV generate horizontal reactions and in addition the windward footings become vertically unloaded whilst the leeward legs becoming vertically loaded, under the storm loading.

Provided the storm footing reactions lie inside or on the boundary of this factored bearing capacity envelope, the foundation will support the loads applied by the vessel on its spudcans. A utilisation of 1.0 is representative of any one footing reaction reaching the boundary of the V-H factored capacity envelope as is the case in Figure 6. Since there are thousands of sets of bearing capacity checks, they are performed using the Python post-processing script with the capacity envelopes programmed in equation form. Spot checks similar to Figure 6 are produced to confirm the script is correctly identifying limiting combinations of wind speed, wave height and waterdepth.

Once the nomograms are produced, benchmarking studies are conducted to compare the results from the nomograms against equivalent SSAs. In some recent comparisons, it was found the nomograms suggested significantly reduced operability compared to the SSAs to the extent that, other than for the simplest of cases, they were unworkable for their original purpose of being used in the Location Approval process.

As previously mentioned, nomograms can only present information for three variables, with all other parameters fixed. The fixed parameters must take quite conservative values for the nomograms to represent a wide range of locations. In this case study, one of the fixed parameters is the foundation model, set to the conservative assumption of a pinned foundation in sand (i.e. the foundation provides no rotational moment or stiffness).

Two examples of changes made in more recent nomogram studies to reduce these conservatisms are now discussed.

One of these was by introducing foundation moment capacity and rotational stiffness into the model. This reduces the applicability of the nomograms to locations were a specified minimum level of spudcan penetration into the seabed is achieved. However, the presence of rotational resistance at the foundations reduces the magnitude of the peak leg bending moments and the axial loading on the jack-up footings caused by environmental loading on the jack-up. With reference to Figure 6, this means that the storm footing reactions do not fan outwards from the stillwater footing reaction as far as they do using the pinned foundation model, and the rig can be shown to be able to resist larger storms.

There can also be a benefit of reducing the dynamic response of the structure and corresponding inertia loads, because the rotational stiffness at spudcan-soil interface reduces the sway natural period of the jack-up, in some cases moving it away from the wave peak periods of excitation.

The second change made in more recent assessments is adopting ISO 19905-01:2016 criteria [2] instead of SNAME, which has the specific benefit of having a foundation bearing capacity envelope that affords greater bearing capacity at the high-end of the loading regime compared to the equivalent SNAME curve. Figure 7 shows the difference between the SNAME and corresponding ISO foundation bearing capacity envelope.

The impact of these two changes (introducing seabed fixity and adopting ISO criteria) are nomograms with much increased operability as is demonstrated graphically in Figure 8. This shows that the acceptable wave height can be nearly doubled in some circumstances, making the new curves better suited for use in Location Approval purposes.

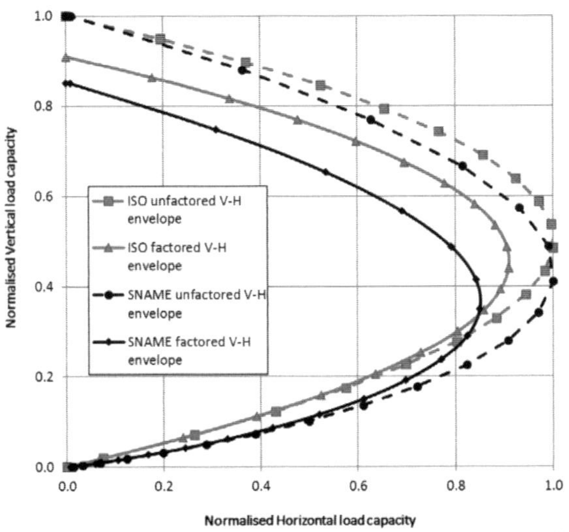

Figure 7 – Comparison between SNAME and ISO sand foundation bearing capacity envelopes

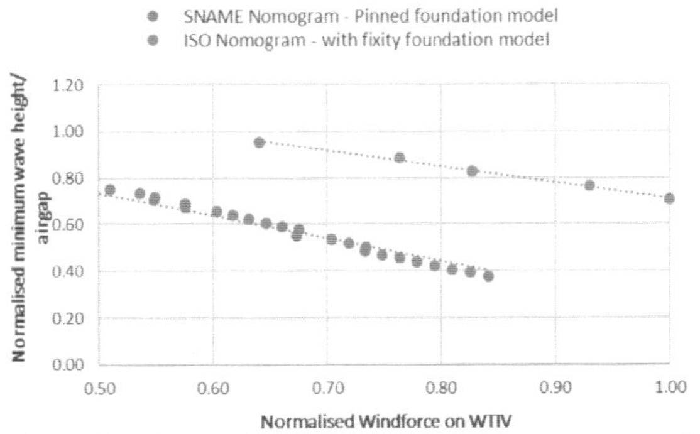

Figure 8 – Comparison between a nomogram generated using original and updated foundation assumption.

7. CONCLUSIONS

This paper has presented examples of the use of Nomogram Operability Curves in the context of jack-ups. It has traced their origins in the Oil & Gas Industry, to their uptake in the Wind Energy sector with specific reference to WTIVs.

There is an understandable drive amongst operators to ensure their WTIVs are used to the maximum safe limits of their operability (c.f. [1]) as windfarms move steadily into deeper water-depths and also harsher weather environments. Operators are also seeking to understand as quickly as possible which locations present the most severe loading of their WTIV's to optimise installation sequences, reduce installation time and therefore reduce cost and improve efficiency.

Nomograms are an attractive tool to help them achieve this, since they are relatively easy to interpret, concise and accessible to the non-specialist user, particularly when packaged into an interactive spreadsheet or computer program.

However, some notable disadvantages of nomograms have also been presented. Due to the simplifications they contain, they cannot in general be expected to demonstrate the same extent of operability as an SSA. When faced with a location for which the nomograms give a "fail" in terms of storm survivability, it is not as easy to refine the analysis and thereby adjust the answer, due to the sheer volume of cases considered in nomograms. Instead of trying to change the nomogram it is generally simpler to run a site specific analysis for the case in question.

Nomograms have a valuable place in allowing a first pass assessment to be undertaken for any new field but they have the disadvantage that they may significantly underestimate the capability of a WTIV at a location. Therefore additional SSAs for problematic locations are still likely to be required if the full capability of the WTIV is to be exploited.

A case study with a focus on foundation bearing capacity checks has been presented showing how the operability of nomograms could be improved significantly by taking care to select appropriate input assumptions, ultimately to avoid multiple conservatisms becoming compounded and resulting in a set of nomograms which is excessively conservative.

8. ACKNOWLEDGEMENTS

Authors would like to acknowledge the input of their colleagues within DNV GL, notably Mark Hayward, Kevin Drake and Richard Stonor.

9. REFERENCES

1. HAYWARD, M.L., 'Going further offshore – assessment of Jack-Up Wind Turbine Installation Vessels', *Design & Construction of Wind Farm Support Vessels*, 30-31 March 2016.
2. ISO 19905-01:2016(E). 'Petroleum and natural gas industries – Site specific assessment of mobile offshore structures, Part 1: Jack-ups', January 2016.
3. SNAME, 'Technical and Research Bulletin 5-5A Recommended Practice for Site Specific Assessment of Mobile Jack-Up Units', First Edition, Rev 3, August 2008.
4. American Institute of Steel Construction, 'Specification for Structural Steel Buildings', March 2005.

10. AUTHORS BIOGRAPHY

Zahidur Rahman holds the current position of Senior Engineer in the Jack-up, Geotechnical and Metocean team at DNV GL. He is responsible for executing and managing projects relating to Location Approval of jack-ups for the oil & gas and offshore wind sectors. His previous experience includes 4 years technical assessment of oil & gas and windfarm installation jack-up units for operating in locations worldwide, as well as delivering technical training courses for jack-ups to rig owners, operators and designers. Prior to that Zahidur worked for 6 years in the Floating Structures department of Noble Denton Consultants Ltd. on Naval Architectural aspects of offshore operations including Heavy Lift Vessel transports, FPSO inclining tests, and mooring analysis & design for offshore floating structures. Zahidur is a Chartered Engineer and Member of RINA, holding Masters degrees from the Universities of Cambridge and Southampton.

Ankor Raithatha manages the Jack-up, Geotechnical and Metocean team at DNV GL, which is based in London and has a long and established history in contributing to industry best practices and providing engineering services for jack-ups for locations worldwide. As well as overseeing the team, Ankor provides consultancy services for jack-up units including site specific assessments and location approvals, advanced structural analysis, technical training courses and intermittent attendance offshore to provide technical operational support. Prior to focusing on jack-ups Ankor worked in the buildings industry, providing structural design and advanced analysis services for high profile architectural projects. Ankor is a Chartered Engineer, a member of the IMechE and he holds both an undergraduate (MEng) degree and a doctorate (DPhil) from the University of Oxford.

INTELLECTUAL PROPERTY IN THE MARINE & WFSV SECTORS - LEGAL ISSUES & PRACTICAL TIPS

R Edson, Withers & Rogers LLP, UK

SUMMARY

The following article first aims to provide a brief introduction to the basics of Intellectual Property rights relevant to the marine sector, for the uninitiated. It further goes on to describe some of the more specific aspects of the international Intellectual Property regime which apply in the marine sector. Some practical examples of types of Intellectual Property used in the marine sector are given and illustrations of the types of commercial intelligence which can be gathered from the published Intellectual Property registers are given, with specific reference to an example from wind farm support vessels.

1. INTRODUCTION

A range of Intellectual Property (IP) rights have long been available to inventors and designers of all manner of technical innovations, allowing them to apply for registered Intellectual Property protection of various kinds, in practically all countries worldwide.

Intellectual Property rights give their proprietor an exclusive right, i.e. the right to exclude (prevent) third parties from taking advantage of the proprietor's Intellectual Property rights for their own commercial gain. This will be well-known concept for those who are familiar with the Intellectual Property systems, but it is an important distinction to understand when beginning to consider IP rights in any depth for the first time. This is because, obtaining a granted patent for a particular invention can give the proprietor of that patent the right to prevent others from using his patented invention. However, patent protection does not give the proprietor the right to use his invention with any immunity from infringing third party rights. Therefore, if an inventor develops an improvement to an already patented product, he may obtain patent protection and thus be the only person who can use that improvement to the original product. However, he would still require a licence from the proprietor of the patent on the product to which he makes an improvement in order to be able to commercially exploit his own 'improvement' invention. A form of shared rights on the improvement therefore exists, where patent rights from both parties are required to exploit the improved invention. This is an important aspect of the Intellectual Property systems, which is intended to enable inventors the freedom to improve upon other inventors' products and to obtain protection for improvements which they have developed. However, this simple act of improving on another's invention should not deprive the earlier inventor of the right to charge a fair license fee to anybody who benefits from improvements upon his original invention. Similarly, if the original inventor wanted to use the second inventor's improvement, that first inventor would also have to reward the second inventor with an appropriate license fee. Managing the risk of infringement of third party rights requires an awareness of what rights exist. It is also necessary to understand what inventions have already been made to assess whether one has truly invented anything new. For this reason, comprehensive IP registers have been developed and are publicly accessible.

The Intellectual Property registration systems have a series of public registers, which are available in electronic on-line format and are freely accessible via a number of websites, such as via the website of the World Intellectual Property Office (WIPO), or via a portal provided by the European Patent Office (EPO), known as Esp@cenet (www.espacenet.com). These portals allow any person the ability to search for patent documents, to download and read those patent documents and to learn from its technical content. In a more legal context, the published documents allow IP attorneys to advise their clients on whether they are at risk of infringing a third party's rights by commercialising a certain invention and whether they appear to have a new and 'patentable' technology. On a more global scale, commercial searching packages are available which allow a more detailed global analysis of patent application data, so that a global view of patent activity in a certain technical or industry area can be carried out. Drilling down to technical details is possible where necessary. These analyses can also provide a more global overview of which companies are filing applications and in what technical areas.

The following sections describe some key issues of the most direct relevance to Intellectual Property in the marine sector, but many of the concepts apply to other sectors as well. For the avoidance of any doubt, the opinions expressed are those of the author alone. Although discussing matters of Law, the information presented should not be relied upon as any form of legal advice. Before taking any action relating to the issues discussed in this paper, direct legal advice should be sought based upon the facts of specific circumstances and this general commentary should not be relied upon.

2. BACKGROUND - TYPES OF IP PROTECTION

In a technical context, there are two main types of Intellectual Property protection. There are rights which exist automatically, the most commonly known being copyright, and there are rights which are obtained through registration systems, known as Registered IP Rights. These include, patents, design and utility models, which provide protection for technical features of products or processes, while trade mark registrations protect brands. The former, protecting technical and aesthetic features of products or processes, are the focus of this discussion and details of the various types of IP are discussed in the following.

2.1 REGISTERED vs UNREGISTERED RIGHTS

2.1 (a) Copyright

Copyright exists automatically as soon as a work is created. It can be enforced against third parties who have been proven to have copied a copyright work. This can require a preregistration in some countries, such as the USA, but in many countries, no registration of copyright is required for enforcement. Copyright does essentially "what it says on the tin", i.e. it protects the author's rights <u>to make copies of</u> the original work. However, key to copyright protection is the requirement to <u>show the copying</u> of the original work with adequate evidence, which can be a difficult and expensive fact to prove, unless clear and incontrovertible evidence of copying is available.

Systems of registered rights generally exist to remove the requirement to show any such copying, making enforcement more straightforward. The first person to file for registered IP protection has the right to enforce that against any infringing third party, regardless of whether the third party independently created the infringing product or not.

Although copyright can be useful in the protection of technical drawings being copied, and for the protection of information contained in the form of written works in, for example, technical manuals, or website material, its application to technical products is often limited by national Laws. This is largely because the duration of copyright protection is 70 years from the death of the author and, for various reasons, which some might consider a little unfair on engineers and designers, this long duration of copyright is generally reserved only for artistic or literary works. If you are an engineer, you are unfortunately not given such long ranging protection. A broad policy reason behind this is to encourage technical inventors to keep inventing to ensure the continued development of new technologies. In essence, after 20 years, any patented technology becomes free for the world to use and if the inventor would like a further monopoly right, then he must invent a new product, or a patentable improvement on an existing one, to guarantee that further monopoly.

In a technical context, the most relevant rights to consider are generally patents, utility models, registered designs, and unregistered design rights.

2.1 (b) Patents

Patents are intended to cover functional aspects of how a technical invention works. An application for a patent consists of a written description of how a particular example of the invention might work, usually accompanied by drawings or diagrams. However, the key part of a patent application is the claims, which are usually found at the end of the published application, or of the granted patent. The claims seek to define, in words, the broadest technical definition of the invention, which includes enough features to differentiate the invention from known solutions to the problem addressed by the patent. The aim to give a broad definition of the key features of the invention, while also covering obvious variations on the main solution devised by the inventor. Patents give a 20 year duration of protection, starting from their filing date. Being a registered right, there is no need show any copying in order to show infringement of a patent. Simply put, if an infringer starts using your technology, after the date of grant of a patent on your invention, then you can pursue them for infringement, whether they have copied you or not. Various remedies for IP infringement are available, including an injunction preventing the third party from continuing their infringing acts, monetary damages, which can be calculated in a number of different ways, and the infringer can also be forced to either destroy or hand over any infringing products in his possession or under his control.

Patents are territorial, and so have to be applied for on a country-by-country basis, although some regional systems exists, for example the European regional patent system, administered by the European Patent Office (EPO), which covers much more the EU, and now has 38 member states, plus additional coverage stretching as far as Morocco and Moldova. Even Cambodia is now on the path to recognising EPO patents as enforceable in its territories from July 2017.

To get a patent granted, first an application is filed, that application is published, then it is examined on its technical merits by the Patent Offices in any countries in which the patent has been applied for, and then once the various legal requirements have been met, then the patent will be granted to the Applicant and they can, after grant, pursue infringers for any of the various remedies for infringement that are available. The key requirement is that the claims in the application must define an invention which is both new compared to previous solutions, and is also non-obvious when compared to known technologies,

2.1 (c) Registered Designs

Registered designs protect **aesthetic features** of products. If a patent protects *"how it works"*, then the design registration protects *"how it looks"*. The details of the registration systems around the world vary, but in essence, a set of drawings or illustrations are registered showing the look of the design for which protection is sought and, after examination, which is at different levels of detail in different countries, the registration is eventually granted. After grant, the registration can be enforced against third parties. The duration of design protection varies from country to country, for example from 14 years from the grant date in the US, to 25 years from the filing date in the EU, subject to renewal fees that are paid periodically, every 5 years in the example of the EU and UK.

2.1 (d) Utility Models

Utility models essentially sit somewhere between patents and registered designs. Utility models look a lot like a patent application and usually have the same detailed description, figures and claims. However, they are cheaper and easier to apply for than patents and they generally have a lower threshold for "obviousness" than patents. They are therefore cheaper and easier to obtain, but the payoff is that you have a slightly shorter duration of generally around 10 years, again varying from slightly country to country. Utility Models are not commonly examined on their technical merits before being granted. That examination will instead happen either when a third party wishes to challenge the registration of the Utility Model on the basis that it is not technically innovative enough, or otherwise when the proprietor seeks to enforce the Utility Model against a third party infringer. Although they are cheaper and simpler rights to obtain, Utility Models can be quite powerful in certain situations and can be a helpful complement to a more broadly defined, longer lasting, patent portfolio. This can be especially useful where an applicant wishes to extend the geographical coverage of their IP portfolio to some additional countries, for a lower price than that of obtaining full patent protection.

2.1 (e) Unregistered Design Rights

Unregistered design rights exist in certain countries and are akin to a copyright-eqsue right for industrial products. The right exists automatically once a design is created. To qualify for this automatic right in a particular country, the designer or proprietor often has to have some link with the country in question. For example, in order to qualify for UK unregistered design right, a design must be owned, licensed or designed by a UK entity, or by an entity in an EU country, or by an entity based in an overseas country which offers a similar reciprocal right to UK entities and designers. UK companies and designers also qualify for an EU-wide design right under EU Design Law. So, for example, a UK employer of a designer will enjoy both a 3 year EU-wide unregistered design right which can be enforced against copiers across the EU. Additionally, under National UK law, a 10 year unregistered design right also exists, enforceable in the UK against copiers of a design created by that UK company. Being unregistered rights, the chain of copying discussed above in relation to copyright must be demonstrated in order to enforce against third parties.

3. SPECIFIC EXCLUSIONS TO MARINE IP PROTECTION

Intellectual Property systems provide a number of exclusions to Intellectual Property infringement. These exclusions are generally provided for policy reasons, in order to allow certain activities to be carried out without fear of infringing IP rights. Private non-commercial use is exempt from infringement. Academic Research is also usually exempt. Some of these exclusions vary country by country. For example exclusions to patent infringement by carrying out medical treatment or surgery exist in Europe, in order to remove from a doctor's mind any concerns about patent infringement at the point of administering treatment. This is not the case in all countries, such as the USA. The European exclusion to patent infringement by treatment or surgery places the concern and any necessary investigations regarding patent infringement with a procurement department well before the point of administration of any treatment, where more considered decisions and investigations can be made over whether to procure a particular medical device or apparatus.

A policy-based exclusion particular to marine IP protection is the 'visiting ships' exclusion, which excludes any ship temporarily visiting national waters from being pursued for patent infringement, subject to certain conditions described in the following section. This exclusion is more uniformly implemented worldwide, due to its presence in an International Convention which has been adopted by a large number of countries worldwide.

3.1 SHIPS TEMPORARILY ENTERING NATIONAL WATERS

The particular exclusion relevant to marine sector is an exclusion to IP infringement for ships (and their on-board equipment) which are temporarily entering the National waters of a particular country, when the ship is registered overseas (the 'visiting ships' exclusion). The policy intention of this exclusion is to prevent patent protection on all or part of a ship from having a restrictive effect on International trade in the mere goods or people which the ship is to deliver to, or take from, a country where patent protection exists. The effect is that, so long as the ship is only temporarily entering the waters of another country, to deliver goods or passengers to or from that country, then the owner or operator of that ship does not need to worry about patent infringement in that country. However, if the ship wishes to carry out operations within that visited country, e.g. moving goods within that country, or wishes to essentially "take up residence" in that country, then the ship would no longer be temporarily entering in the eyes of the patent Law and could therefore be vulnerable to claims for patent infringement. If a vessel is registered in a particular "home", or "flag" state where relevant patent protection exists, then the vessel can be pursued (and possibly impounded) for patent infringement when docked in its home or flag state (if it ever is).

Where a wind farm support vessel is carrying out support operations to wind turbines in the national waters where relevant patent protection exists, the question that arises is that of whether such a vessel, if it where registered in a different country, would be liable to claims for patent infringement in the country in which it operates. That question is considered in further detail in later sections. First, a brief review of the wording under some different national, regional and international Law is set out in the following sections.

National and Regional laws have defined this "visiting ships" exclusion in slightly different words, which can result in different subtleties in interpretation of the exclusion for different rights in different countries.

3.1 (a) In the Paris Convention for the Protection of Industrial Property (The Paris Convention) [1]:

The Paris Convention has existed since 1883 and most countries of the world have now joined. The convention has a total of 177 contracting parties, the most recent addition being Afghanistan, which will formally join on 14 May 2017. This convention therefore forms a core piece of IP Law governing all of its contracting states. Its provisions therefore apply to all 177 of its signatory states.

Acts excluded from patent infringement by the Paris Convention are:

*"(i) the use on board vessels of other countries of the Union of devices forming the subject of his patent in the **body of the vessel, in the machinery, tackle, gear and other accessories**, when such vessels **temporarily or accidentally** enter the waters of the said country, provided that such devices are used there **exclusively for the needs of the vessel**;"*

The highlighted parts in particular mean that patent infringement for vessels temporarily or accidentally entering the waters of a particular country is excluded, provided that the devices in question are used there "*exclusively for the needs of the vessel*".

3.1 (b) In the UK Patents Act 1977

The UK Patents Act 1977 [2] defines national Patent Law in the UK, and incorporates, in very slightly different wording, the requirements of the Paris Convention defined above. Again, the key requirements are that the use must be *"explicitly for the needs of a relevant ship"*, and it applies where the ship has *"temporarily or accidently entered the internal or territorial waters of the United Kingdom"*.

Art 60 (5) (d):

"(5) An act which, apart from this subsection, would constitute an infringement of a patent for an invention shall not do so if—"

...

*"(d) it consists of the use, **exclusively for the needs of a relevant ship**, of a product or process **in the body of such a ship or in its machinery, tackle, apparatus or other accessories**, in a case where the ship has **temporarily or accidentally** entered the internal or territorial waters of the United Kingdom;"*

3.1 (c) In the Community Design Regulation, (*"The CDR"*) [3]

The CDR defines how Registered European Community Designs cannot be enforced against ships and aircraft which are registered in a third country when they temporarily enter the territory of the (European) Community. The CRD also makes it explicit that the importation of spare parts for the ship or aircraft, and the execution of repairs on such a ship or aircraft, would not constitute infringement of European Community Design Rights.

Extract from the CDR:

"the rights conferred by a Community design shall not be exercised in respect of:

(a) the equipment on ships and aircraft registered in a third country when these temporarily enter the territory of the Community;

(b) the importation in the Community of spare parts and accessories for the purpose of repairing such craft;

(c) the execution of repairs on such craft."

These exclusions are again clearly implemented to avoid anyone entering EU waters temporarily and needing to repair their vessel being 'held hostage' by a design right holder in the EU, who could, in theory, otherwise obtain an injunction to prevent them from repairing their vessel by importing infringing products (spare parts) into the

EU. Such a restriction on and free movement of goods and passenger vessels is therefore not permitted.

3.2 GENERAL COMMENTS ON THE EXCLUSIONS

It is worth noting what activities do not fall within this 'visiting ships' exclusion. Firstly, patent protection in general does cover all acts of making, using, offering for sale, stocking for sale, or offering for sale, any patented product in a patent-protected territory. The 'visiting ships' exclusion would not cover any shipyard building an infringing vessel, or equipping a vessel with infringing machinery, in a patent protected territory. IP rights would remain enforceable against those activities. Further, anyone selling or offering for sale, stocking or importing an infringing vessel in a territory where relevant IP protection is in place would not fall under the exclusion and could therefore also be pursued for patent infringement.

As regards wind farm support vessels in particular, the question then arises over whether a vessel which is registered in a foreign state might, for example, infringe patent rights, covering some of its general structure or equipment, by travelling each day to UK waters to operate there. For example, the vessel might be registered in the Cayman Islands, and may travel to the UK waters from France, Belgium or the Netherlands each day. For the purpose of the example, we would assume that no relevant patent exists in those countries and only UK patent protection (or UK-only design protection) has been obtained.

The primary question revolves around whether the vessel is "temporarily" entering the UK waters, where the IP protection exists, within the definition of "temporarily" which has developed under the relevant Case Law.

This type of question has already been considered by the UK Patents Court and the Court of Appeal (*Stena Line Aktiebolag vs Irish Ferries Limited, [2003] EWCA Civ 66 [4], 'the Stena Line Case'*). In the Stena Line case, the UK Patents Court considered the meaning and effects of the defence for ships temporarily entering UK waters. The judgement made by the Patents Court in this case was upheld by the Court of Appeal.

The case concerned a high-speed catamaran which was used to provide a regular ferry service between Eire and the UK. Three or four crossings were being made each day. The vessel's home port was in Dublin, but it spent around 3 hours in UK territorial waters on each crossing. The superstructure of the vessel was found to fall within the scope of protection of a patent held by the claimant, Stena Line. Stena Line then argued that the vessel's regular and frequent crossings into the UK took its activities outside of the scope of the defence of section 60(5) (d) of the UK Patents Act, because "temporarily" should be interpreted as "on isolated occasions or casually". The argument being that regular scheduled commercial visits were not "temporary or accidental". In this case, the court rejected the argument, and stated that the primary purpose of the word "temporarily" was to distinguish between vessels which were engaged essentially *internal* operations and those which *travel between countries* for the purpose of their operation. A further relevant comment on this case is where the judges had referred to the "*travaux preparatoires*" for the Paris Convention (the original piece of International Law defining this exclusion discussed above) and also a German judgement, of the district court of Hamburg in a case known as *"Rolltraller"*, (reported in GRUR in 1973 at page 703). The Patents Court judge (whose decision was upheld by the Court of Appeal) explained:

"it seems to me that [the 'visiting ships' exclusions] are concerned to ensure that trade and carriage of persons between countries are not hindered by patent rights applying to the means of transport. Whether one considers this a ship, aeroplane or lorry, each is made up of large numbers of parts and accessories. Anyone of them could be the subject of patent protection. The owner of a patent to the vehicle itself [or] any one of its numerous parts or accessories could immobilise the vehicle by infringement proceedings and thereby interfere with International trade or the International movement of people, at least unless and until an alternative means of transport can be put in place. Whether the means of transport takes the form of lorries, planes or ships, the underlying principle is that the normal operation of the vehicle is to be taken out of the scope of patent infringement. If the vehicle is used for an infringing purpose, then patent rights can be evoked, but not if the patent covers the fabric of the vehicle or a process which enables the vehicle to function".

"If that is correct, then [the 'visiting ships' exclusion] comes into play to protect the vehicle in so far it is engaged in inter-state passage. The word "temporarily" should be construed in that context. Its primary purpose is to distinguish between vehicles which were engaged essentially in internal operations and those which travel between countries. For that reason, I accept Mr Miller's argument, which is consistent with a decision, that questions frequency, persistency and regularity have little to do with determining whether a vessel is temporarily within the territorial waters of the United Kingdom. Temporarily means "for a limited period of time"."

This reasoning is interesting in relation to different cases in which a wind farm support vessel might enter the UK waters. It may merely enter to deliver crew to a wind turbine, in which one might conclude that the situation is precisely the same as that of the ferry considered in the above case. In a one-stop case, this would be a similar situation to the ferry in the Stena Line case. However, the vessel may carry out repeated pick-ups and drop-offs in a day on various wind turbines within UK waters. This may move it more towards the kind of 'internal operations' discussed above in the Stena Line case, which may move it out of the exclusion. If transferring people between wind turbines during its visit, then the vessel is more likely to fall outside the exclusion and to be being used for an 'infringing purpose' as referenced by the judge above.

Further, the vessel may include particular equipment or apparatus which is specifically configured for attaching to, or interacting with, a wind turbine to perform maintenance operations. The Paris Convention and the UK Patents Act both require that the use of equipment must be **"exclusively for the needs of"** the relevant ship or vessel. Equipment for transferring people or equipment on or off the vessel, or for attaching the vessel to a wind turbine for maintenance operations needed by the turbine and not by the vessel would therefore arguably not fall within the 'visiting ships' exclusions, since it is perhaps not 'exclusively for the needs of the vessel'.

Patents can also cover methods of maintenance of a piece of apparatus. Therefore, a patent drafted with appropriate claims, covering a method of maintenance of the wind farm or wind turbine, where the apparatus includes some bits of technology attached to the wind turbine, and other corresponding parts on the vessel, then the exclusion would arguably not apply. The use in UK waters would more likely be for an 'infringing purpose' as discussed in the Stena Line case.

Under European Design Law according to the CDR referred to in 3.1(c) above, the Registered Design Rights in any parts of the vessel seem to be more clearly covered by the exclusion defined in the CDR, where the vessel temporarily enters European waters from International or non-EU National waters of another country. This is because the CDR clearly states that:

"the rights conferred by a Community design shall not be exercised in respect of:

(a) the equipment on ships and aircraft registered in a third country when these temporarily enter the territory of the Community"

No *"exclusively for the needs of the vessel"* requirement is present in the CDR. Note that an EU registered design would cover the UK, France, Netherlands and Belgium in the above example, so a more relevant example to consider in that respect might be a vessel visiting Spanish waters from North Africa, for example.

IP attorneys drafting patent applications protecting equipment on ships or vessels intended for use in wind farm support operations, or indeed in any operations at sea, will therefore need to be careful to claim coverage which covers methods which extend as far beyond the needs of the vessel as possible, covering technical equipment which crosses the boundary between the vessel itself and the actual wind turbine as far as possible, and any claiming any activities not directly relating to inter-state-transport, to try to navigate around these exclusions.

4. GEOGRAPHICAL STRATEGIES FOR MARINE IP PROTECTION

Taking into account the above exclusions, anyone wishing to protect the equipment which will be merely located on vessels for the transfer of goods or persons must therefore be sure to gain coverage in countries where the vessels might be equipped or constructed. In this case, even if the transfer of people or equipment falls under the exclusion described above, the actual act of constructing the vessel will still be covered by patent protection if it is in place in the country where the vessel is built. The person or entity which constructed or sold the vessel in a country where IP rights subsist, could be sued for retrospective damages, generally for any acts up to 6 years in the past in most countries. Therefore, even if infringement is not detected and halted during the actual build, once the craft is launched and operational, a notional licence fee, full damages, or all profits obtained from that sale or construction of the vessel could be sought from the infringing builder or vendor via the IP courts.

Further, if protection is primarily sought in the area of operation of a vessel, for example in the UK, it should be taken into consideration what other waters the vessel might travel *from* in order to operate in the UK. For example, by gaining patent coverage in France, Belgium the Netherlands, Germany and/or Norway would mean that any vessel travelling regularly from those other countries whilst being resident there, would be likely to be caught for infringement in its "home base", even if not caught during operations in UK waters. Patent coverage may therefore be sought via the European Patent Office (EPO) to cover coastal European countries. This can be complemented by a design registration, covering the whole of the EU. This can also be further complemented by coverage in other major ship-building countries, such as Turkey (also obtainable via the EPO), China, Korea and the US, as a few examples.

5. SOME PRACTICAL EXAMPLES

A brief discussion of real examples of IP protection in the wind farm support vessel sector now follows, in order to put some of the above into practical context.

5.1 PRACTICAL EXAMPLES

5.1 (a) Patent GB2480408 B2 : Transfer apparatus with compensation means for vessel movement

Figure 1 below shows an apparatus intended for assisting with the delivery of persons to a wind turbine, and including a compensation system to help maintain the platform 10' in a relatively steady position with respect to the wind turbine 100. Pulley wheels 32' engage with the runners 102 and by a combination of pressure against the runners, and sensors in the runners, the system can help actively move the gangway 10' to compensate for heave and roll in the vessel 1'.

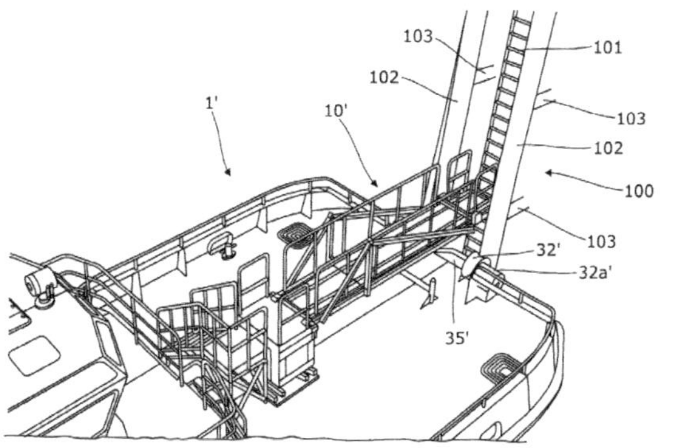

Figure 1 - Drawing from patent GB2480408

This is merely one example of a patent application in the sector, filed by BMT Nigel Gee Limited, and this illustrates the types of technology that can be protected in this sector.

5.2 Patent Classification Searching

To assist with navigating the millions of patent documents which have been filed and published over the years, patent applications are "classified" by the national Patent Offices, based upon what type of technology they relate to and what technology the claims seek to cover. These classifications cover a wide range of technologies and implementations, and involve sub-classifications which allow those searching for patent documents to drill down to a significant level of detail.

For example, the above application is classified in the International patent classification code B63B27/14, defined as follows:

Section B: Performing Operations;
B63 - Ships Or Other Waterborne Vessels; Equipment For Shipping;
B63B27 - Arrangement of ship-based loading or unloading equipment for cargo or passengers
- B63B27/14:... of ramps, gangways or outboard ladders

As is evident from the text of the above classification, it is not limited to wind farm support operations, rather, it generally relates to apparatus for the transfer and unloading of people and equipment.

With a more detailed commercial searching package, we can carry out some statistical analysis to consider which companies are the most active in this classification area and this reveals the top 10 patent applicants in this particular technical classification, by patent family, as shown in Figure 2.

(Note: A "patent family" is, generally speaking, a set of patent applications filed in a list of different countries by an applicant to over the same invention. They are usually based upon one first patent application filed in a chosen country.)

We may however be more interested in that technology in applications which refer to wind turbines or wind farms. We can then add a key word limitation to "wind farm" or "wind turbine" to see what proportion of those applications make any specific mention of this type of subject matter. This doesn't mean that the remaining applications could not be broad enough to cover operations in a wind farm context, but where applications are specifically addressing operations in wind farm support, they are likely to mention these words, and so this key word limitation can be useful as an indicator of patent activity in this area.

This reveals a reduced overall number of applications directly relevant to wind farms, and we can also look at the number of applications filed over time in this context. This shown in Figure 3.

As can be seen from the graph of Figure 3, there is a particular spike in the number of applications filed in 2011 to 2013, and as those applications gradually work through the patent application examination systems, the number of patents granted based on those applications follows, with a slight time-lag.

We can also consider the wind-farm limited data set by applicant, to view how many applications and granted patents each applicant has obtained in this technical area. This is illustrated in Figure 4.

This is just one example of the type of investigation that can be carried out in order to gain commercial intelligence on what companies are operating in a particular area. Once relevant patent publications are identified, they can be studied in detail to see what technologies those companies are developing and patent-protecting. Further, applications are sometimes abandoned before grant, or abandoned after grant. Patent renewal fees are payable every year in most countries, so a company may abandon them once they are no longer of interest. If a company abandons a particular application, then that could be an indication that either technology was not considered commercially successful, or otherwise, if the application was filed before sea trials and testing had taken place, abandonment may show that the technology had proven impractical in operation, for either technical or economic reasons.

6. CONCLUSIONS

This paper has set out to give an introduction to the basics of IP, the peculiarities of the IP systems as they relate to marine products and marine operations, and some illustrious of practical examples, as well as the type of commercial intelligence that can be obtained from the public information on the patent registers worldwide.

It is hoped that this gives both technical innovators, and operators, an insight into the types of IP issues which it is worth considering when devising new technologies and incorporating them into marine operations. Particular points have also been raised, to consider when developing appropriate IP protection and selecting an IP attorney who understands the subtly of these systems and can give proactive advice on how to work around the associated pitfalls.

7. REFERENCES

1. Paris Convention for the Protection of Industrial Property,
of March 20, 1883,
as revised:
at Brussels on December 14, 1900,
at Washington on June 2, 1911,
at The Hague on November 6, 1925,
at London on June 2, 1934,
at Lisbon on October 31, 1958,
and at Stockholm on July 14, 1967,
and as amended on September 28, 1979.

2. The UK Patents Act, 29th July 1977, as amended.

3. COUNCIL REGULATION (EC) No 6/2002 of 12 December 2001 on Community designs

4. Stena Rederi Aktiebolag and Stena Line Aktiebolag vs Irish Ferries Limited, [2003] EWCA Civ 66, Case No: A3/2002/1068; 1104

8. AUTHOR'S BIOGRAPHY

Russell Edson is a Chartered UK & European Patent Attorney and Design Attorney at Withers & Rogers LLP. Russell specialises in Intellectual Property relating to mechanical technology and design rights. He is experienced in patents relating to advanced planing hull technology, advanced power trains and marine turbines. On the designs side, Russell manages multi-national international design filing strategies and advises on infringement issues, coordinating both patent and design filing programmes in parallel. He also advises on managing IP rights for multiple stakeholders in a relationship, such as between ship yards, naval architects, designers and clients.

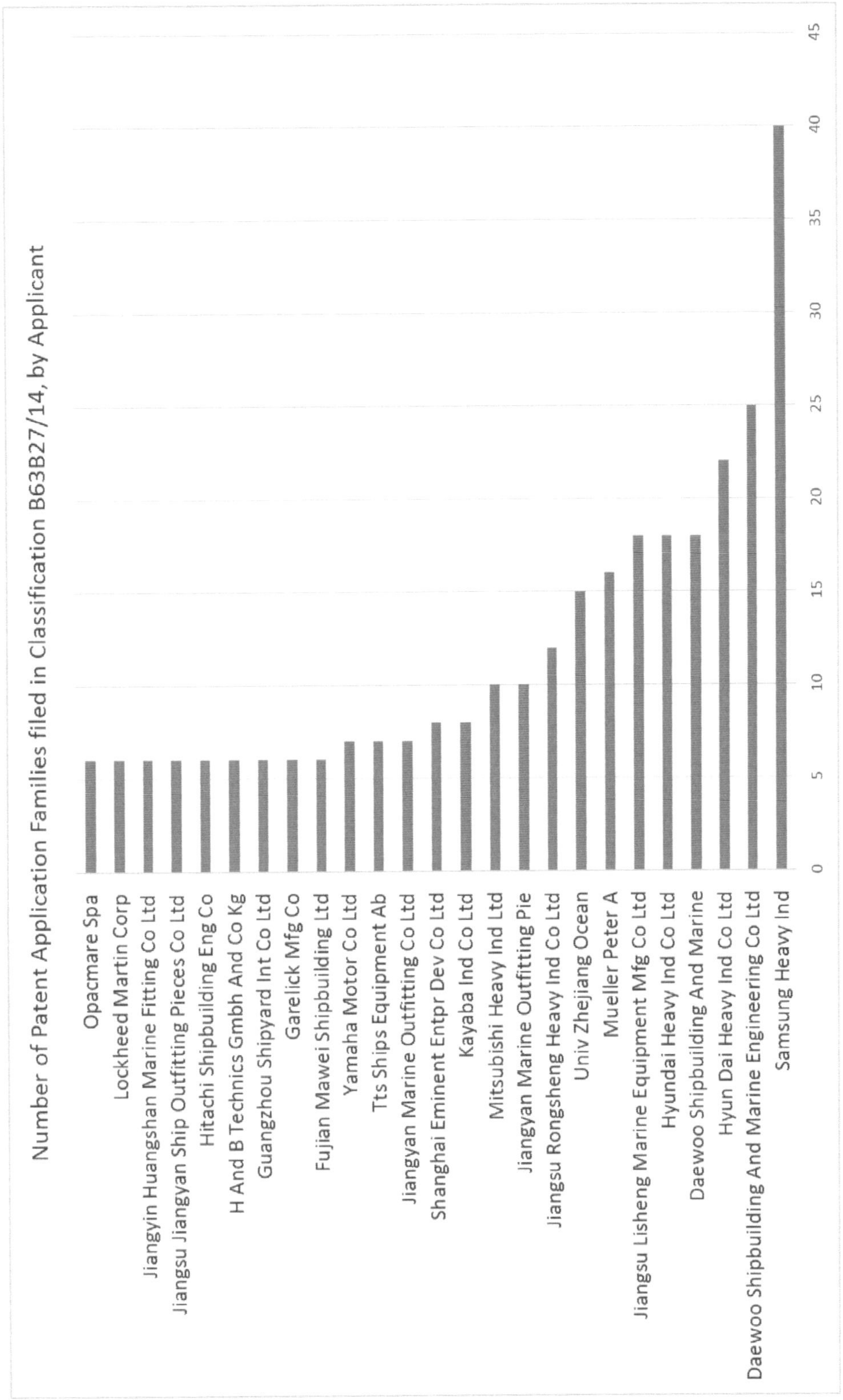

Figure 2 - Number of Patent Application Families filed in Classification B63B27/14, top-10 Applicants

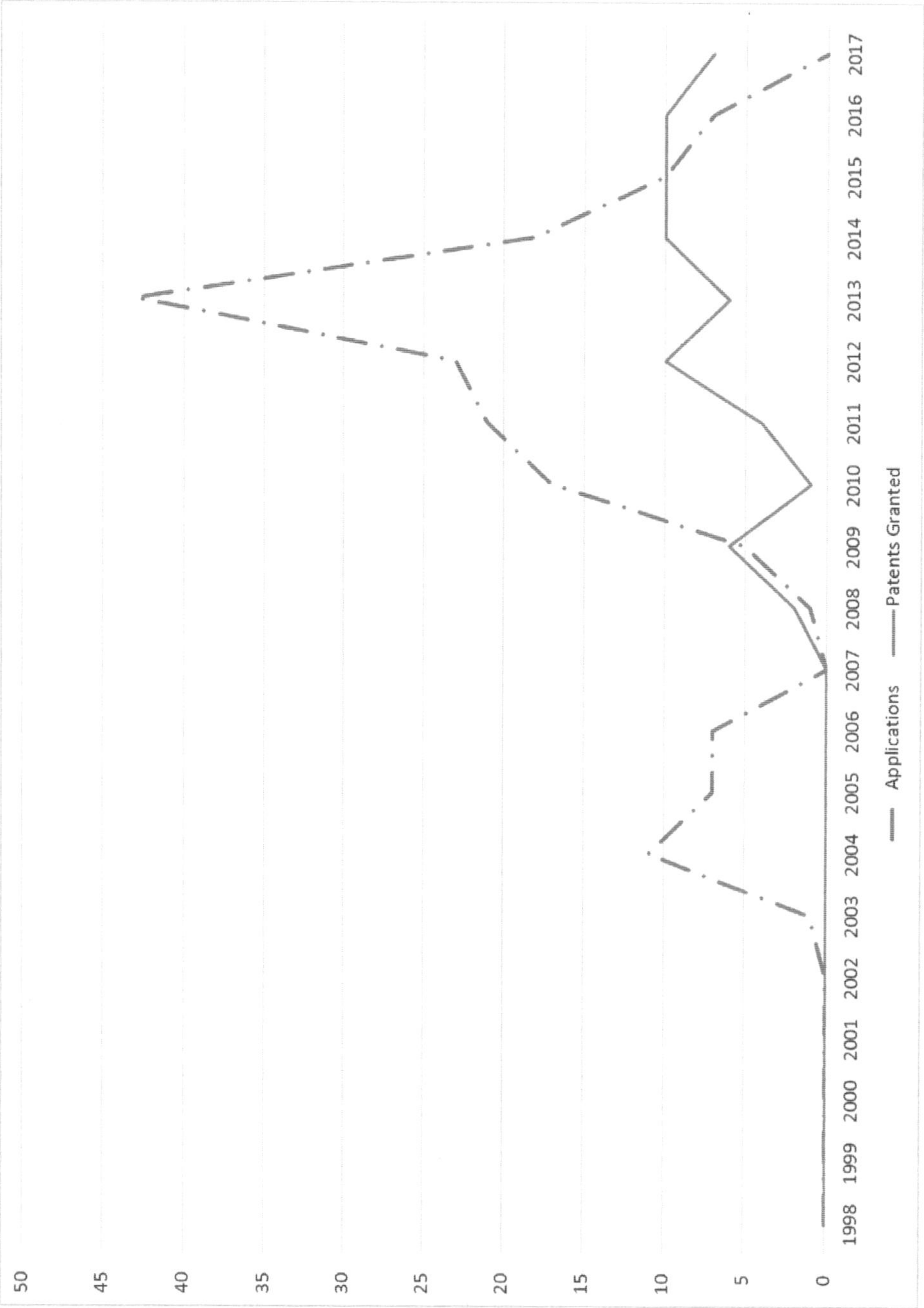

Figure 3 - Number of Patent Applications filed, and granted, in Classification B63B27/14, by year, limited to documents mentioning "wind turbine" or "wind farm"

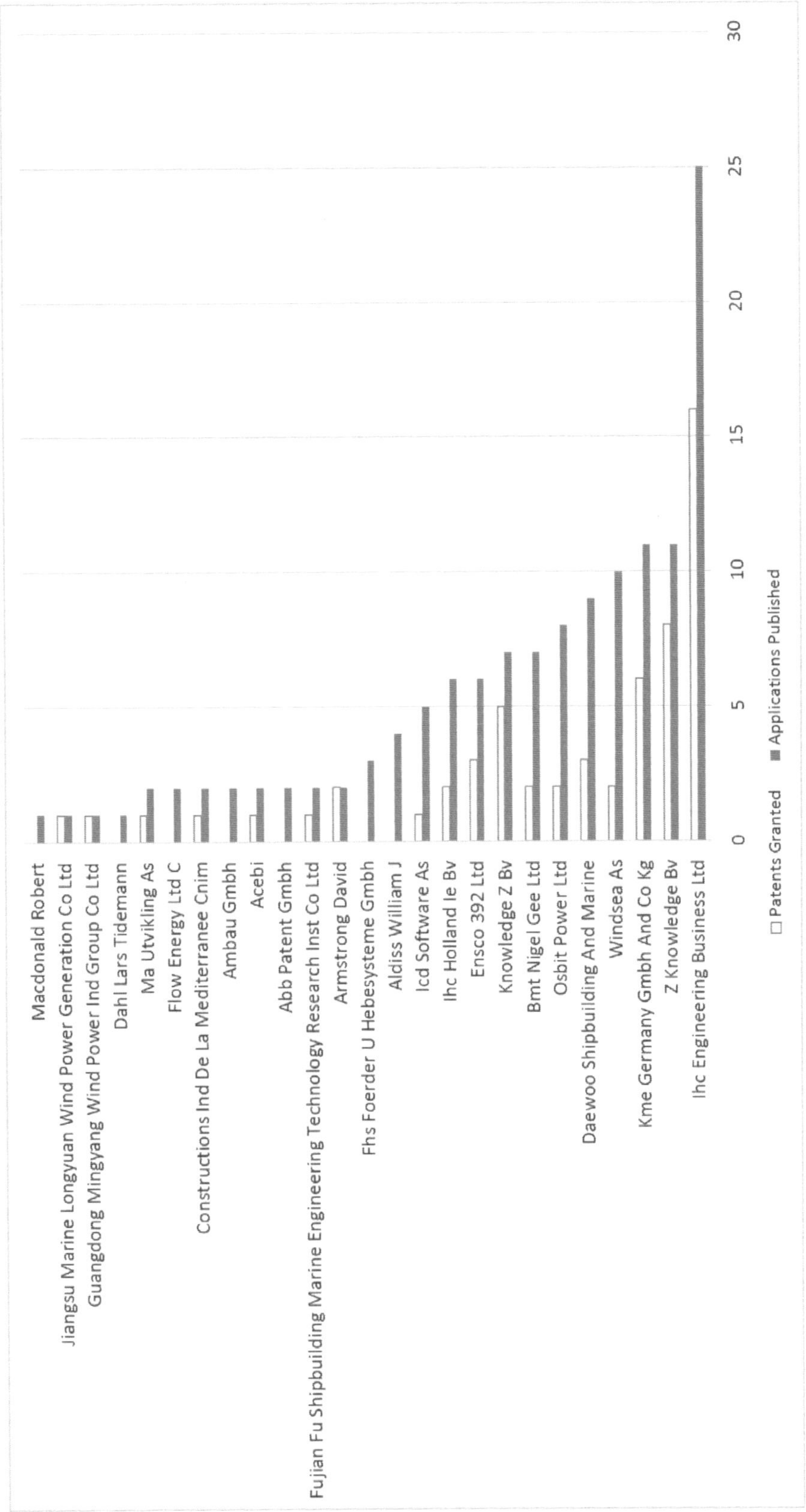

Figure 4 - Number of Patent Applications filed, and granted, in Classification B63B27/14, by Applicant, limited to documents mentioning "wind turbine" or "wind farm"

THE SWATH CTV – FINALLY COMING OF AGE

J.Kecsmar, Ad Hoc Marine Designs Ltd., UK
J. William McFann, Island Engineering Inc., USA
Capt. David Low, Maritime Craft Service (Clyde) Ltd., UK

ABSTRACT

Conventional CTVs have been restrained by legislation, to be below 24m Loadline rules to save costs and further hindered by having to be below 12 passengers otherwise being faced with "big ship" rules. However, whilst these challenges are now being overcome with the new HSC-OSC rule to allow for up to 36 passengers the most difficult challenge remains, that of small boats going out in increasing sea state. All vessels that are in the sub 24m and up to 30-40m exhibit similar natural periods of motion and consequently have similar seakeeping characteristics. The round three windfarms are exposed to Hs=3.0m and existing conventional vessels are not suitable in terms of seakeeping for low motions and the ability to safely transfer technicians at the tower in the increasing sea state, at least not without the aid of a highly damped transfer system. Why focus on the damped transfer gangways alone when the most obvious solution is to have the whole vessel highly damped. If the vessel is highly damped to begin with seakeeping and transfers become significantly improved and safer. The 26m Typhoon Class Swath is such a vessel and has proven to outperform expectations with transfers in sea states beyond any conventional vessel of the same size. Has the once considered "black sheep of the family" type of vessel, a Swath, finally come of age?

NOMENCLATURE

Hs	significant wave height (m)
a	hull added mass and inertia coefficient
b	hull damping coefficient
c	hull restoring force coefficient
$k_{xx,\, yy}$	radius of gyration (m)
g	gravitational acceleration 9.81ms^{-2}
GM_T	transverse metacentric height (m)
GM_L	longitudinal metacentric height (m)
ρ	density of fluid (kg.m^{-s})
Δ	displacement (tonne)
A_W	waterplane area (m)
ω	wave frequency (rad)
ω_e	encounter frequency (rad)
φ	heading
V_s	vessel speed (ms^{-1})
a_{cg}	vertical acceleration (ms^{-2})
ζ	amplitude (m)
T	period (s)

1. INTRODUCTION

The UK windfarm industry has been growing steadily over the past decade. The first large-scale offshore wind farm in the UK, North Hoyle, was commissioned in December 2003, and the second, Scroby Sands, one year later in December 2004, followed by the UK's then-largest offshore wind farm, the 90MW Kentish Flats in

2005. This spawned the current range of workboats which are all catamaran vessels of the order of 12-24m in length (Lwl) with a passenger capacity of up to a maximum of 12 – now termed technicians or industrial personnel. Access to the turbine tower is from shore via these small workboats which run out to the towers and these technicians step across the bow onto a ladder. They operate in the inshore side of the farms of what is termed "round one" and run in generally benign sea conditions but in winter months and when the weather is inclement often have to work in sea conditions of up to 1.5m significant wave height. These first generation catamarans fulfilled the role very well.

However as the windfarms become numerous the role of the catamaran began to evolve from a simple point A to B vessel into a multitasking vessel. With requirements of carrying spare generators on the deck and requiring heavy duty cranes for lifting heavy weights up to the towers etc. As time progressed the round one farms also evolved into round two and subsequently the larger powering generating turbines began the slow migration from inshore to offshore. This is where the technical challenges slowly began to test the limits of the existing designs on two fronts, the first is the seakeeping of the vessel which leads to fatigue of the crew and technicians and secondly the safe transfer of the technicians in the increasing sea state. This paper focuses upon the later aspect whilst noting it is inextricably linked to the former, that of seakeeping.

2. TRANSFER OF TECHNICIANS

A typical transfer of the technicians is shown in Fig.1. The vessel pushes up against two vertical poles either side of the ladder on the turbine tower with sufficient force to hold the bow of the vessel onto the tower, which then allows the technician to step from the foredeck of the vessel onto the tower's ladder.

It can be seen that the transfer shown in Fig.1 is in relatively clam conditions. With such a slight sea state the transfer is easy and conducted in safety. As the number of farms increase and the further offshore they are sited brings with it an increase in the sea states these CTVs are exposed to, not just during the hard winter months but annually. As such it becomes increasingly challenging for the Master of the vessel to perform this operation safely. With the bow of the vessel rising and falling as much as 2.0-3.0m in inclement weather it is clear that the transfer of technicians requires a method or system that is controlled and safe. A method that is less hit and miss and guess work, but one that provides a degree of control and assurance for both the Masters, and the technician that is stepping across from the heaving bow to the turbine ladder.

2.1 Methods of Transfer

An example is shown in Fig.2, which uses a heave compensated gangway. Whilst these types of transfer gangway assist in extending the sea state limits for safe transfers, it can be seen that the whole of foredeck is now taken up by the gangway system. This is a departure from the now multi versatile CTVs where a large open foredeck is the norm.

As such it is hard to image how such units could be used on anything other than a much larger vessel. Since the now cluttered foredeck with cranes and containers etc limits the vessel with such a gangway from the now common multitasking vessel to that of a pure technicians carrying CTV.

Alternative heave compensated gangways have been successfully exploited on previously designed crew transfer vessels in the past. An example is shown on the FBM Designed and built Swath in 1997 [1] that uses a simple single cylinder to damp out vertical heave motions, shown in Fig.3. The gangway, or brow, is located on the side of the vessel for beam to beam passenger transfer from ship to ship. Being located on the beam simplifies the mechanism considerably, since the only motion to account for as the vessel pitches or heaves or rolls, the displacement that is experienced at the side, is purely vertical.

A modified and updated version was designed by Ad Hoc Marine Designs in 2008 for use on the foredeck of another Swath shown in Fig.4. Being on the foredeck increased the complexity of the system as well as the cost. However it was still a compact unit.

The gangways, or brows, shown in Fig's.3 and 4 have been successful principally owing to the benign motions of the vessel it is sited on – this is the principal reason for such small compact units compared to those on offer in today's market. Had these systems been employed on conventional multihulls, it is highly likely that major modifications would have been required. This was most certainly true in the case of the Autobrow, designed by Ad Hoc Marine deigns as part of the Carbon Trust competition for methods of safer transfers. The design was one of the 13 eventual winners, as shown in Fig. 5.

The size of the unit is clearly much larger and the hydraulics required resulted with much higher power requirements. The reasons for the difference between the two, one on a Swath and the other on a catamaran are simple naval architecture and hydrodynamics of the vessels they are sited on. The vessels shown in Fig.1 and 2 are typical CTVs in the 20-24m size. These vessels all exhibit very similar natural periods of motion despite their varying hull shapes; the displacement and waterplane area ratios are very similar.

2.2 Basic Requirements

Taking the typical 20-24m catamaran as an example, what would be a typical pitch period and heave period? For this vessel, empirical evidence (and from the author's own database from sea trials of previous vessels) suggests approximately 2.5 seconds for both pitch and heave. Given this the range of motions and this forces that any gangway or brow is required to compensate for can be calculated. For the catamaran CTV the tangential force when pitching is calculated as 1.2kN the force normal to this as 7.7kN and the vertical heave as 25.6kN, assuming a range of motion of 2.50m or ± 1.25m. A Swath of the same size has a pitch period of 6 seconds and heave period of 5 seconds. Assuming everything else remains the same the tangential force reduces to 0.2kN (83% less) normal to this of 1.3kN (83% less) and the heave to 13.7kN (46% less). It is very obvious that the change in natural periods of the vessel has a significant effect on the forces experienced and hence the power requirements for the damping system that is utilised. And this leads to large cumbersome units like those shown in Fig.2 and Fig.5.

The reason for the big differences in natural periods, and hence forces, is purely hydrodynamic. All objects have a natural frequency of vibration, and this is true of any object be it car or boat. If the natural frequency coincides with the same encounter frequency of each wave this creates the very unpleasant motions called resonance. Therefore to understand the effects of waves on a boat is to understand the natural frequency and what influences them. The period of motion is given by the following expression:

$$a.\frac{d^2\theta}{dt^2} + b.\frac{d\theta}{dt} + c.\theta = \text{Forcing Moment}$$

Where a, b, and c are hull coefficients. The forcing moment is the wave. When this is set to zero, it provides the ability to find the natural frequency of motion.

Rolling Period $\quad T = \frac{2\pi.k_{xx}}{\sqrt{(g.GM_T)}}$

Pitching Period $\quad T = \frac{2\pi.k_{yy}}{\sqrt{(g.GM_L)}}$

Heaving Period $T = 2\pi.\sqrt{\frac{(\rho.\Delta)}{(g.A_W)}}$

What is very obvious to a naval architect is that all three natural periods of motion are influenced by the hull shape. The periods of: roll - GM_T, pitching - GM_L and in heave are all dictated by the waterplane area A_W. Also what we can see is that to increase the natural periods we must either i) increase the displacement or ii) reduce the waterplane area or a combination of the two.

All small vessels beit a monohull or catamaran in the 20-40m range have natural periods of roll, pitch and heave in the order of 2.5 to 4.0 seconds. Thus, the displacement and the size of the waterplane area result in these short periods of motion. This brings us to the influence on waves and their encounter.

The expression that defines how this speed and distance between wave crests affects the encounter period is given by:

$$\omega_e = \omega\left[1 + \frac{\omega V_s}{g} cos\varphi\right]$$

From this expression we can calculate the frequency of encounter knowing basic characteristics of a wave, such as its: period, length and speed, as well as the heading, φ, and speed of the vessel, V_s. Thus for a given wave we can plot how the wave can have an influence on the vessel when running at varying speeds and direction. Typical significant wave height (Hs) values in the North Sea location that are in the round three zone are shown and with corresponding wave heights shown in Fig.6 [2].

We can see that in the summer months the dominate wave heights are in the 1.0-1.50m, but in the winter months 2.0-2.5m dominate. This has been previously highlighted for farms that are up to 300km offshore, with Hs=3.0 being a concern. If we now use this wave data we can plot a speed of vessel versus encounter period, and using a nominal Hs=2.0m and assuming head seas, as shown in Fig.7. Also shown shaded is the typical natural frequency of vessels in the 10-40m range, that of 2.5 -4.0 seconds. What is very obvious is that for a wave of Hs=2.0m for speeds under 30knots the vessel shall have encounter periods matching the vessel natural periods of motion. This means resonance and uncomfortable motions. The only solution is to i) reduce speed to reduce the accelerations being experienced or ii) power through past 30knots.

The accelerations that are experienced owing to resonance of the vessel in the sea state can be calculated knowing that $a_{cg} = \omega_e^2.\zeta$, which means the encounter period squared times the amplitude of the motion. The goal being to reduce the encounter period and the amplitude of the resonance as this leads to lower vertical accelerations. So going faster is a solution, but the vertical accelerations can eventually become unacceptable owing to the high encounter frequency. To put this into a bit more context, the period of motion is given by the simple equation:

$T = 2\pi/\omega$

So, if the encounter period is 10 seconds, this gives ω_e as 0.628 radians. If the period is halved to 5 seconds, it is 1.257 radians and halved again to 2.5 seconds gives 2.513 radians. It becomes clear that for a constant wave height, if ω_e^2 changes the accelerations change too. So with T = 10 seconds $0.628^2 = 0.39$, whereas $2.513^2 = 6.32$, which is more than 16 times greater, i.e. higher accelerations. Clearly demonstrating the link between encounter frequency and accelerations experienced. Thus, increasing the encounter period for a given constant wave amplitude, the resulting change in motions lead to lower vertical accelerations. This is easily achieved by simple manipulation of a Swath hull form, but for a conventional vessel there is very little room for manoeuvring in terms of how much can really be changed.

Since the vessel must go over 30knots to move out of the resonance zone of encounters which inevitably means larger engines and larger jets. This adds to the lightship weight, not ideal for the small craft as this increases capital cost, and consequently making the vessel's daily charter rate higher! It also decreases the length-displacement ratio (LD). There is very clear evidence of decreasing LD ratio increasing accelerations. In other words, the existing CTVs of small catamaran hull configuration are not ideally suited for the round three farms, based upon the simple hydrodynamics of their hull forms.

3. CHANGE OF HULL FORM

When the CTV is running from the port to the tower, it is assumed to be running at its service speed; in this case in the 25-30knots region. As can be seen in Fig.7, with a Hs=2.0m head sea the beginning of the resonance and zone of uncomfortable motions is already set at these speeds; ensuring an encounter outside the 2.5-4.0 seconds is required. A typical fast ferry catamaran that is 30m, like that of RedJet 3, that runs between Cowes on the Isle of Wight to Southampton has a roll period of 2.7 seconds. The LD ratio of this catamaran hull form is suitable for fast ferries and much higher than that of a CTV. However if the vessel were to be made to carrying more payload for the outer farms and remain the same size thereby lowering the LD ratio of the hull form considerably, the motions would not be improved, since the WPA would increase to provide the more buoyant hulls required, thus increasing the GM_T and hence remaining a short period of roll. Changing this roll period is extremely difficult with a conventional hull form, as the resulting displacement required, keeping all other hull attributes the same, requires a change from 90 tonne to 580 tonnes!

3.1 Appendage Options

Other than major changes in the hull form, would adding appendages help the case to continue with a catamaran hull form? Appendages are the "b" coefficient in the motions given in 2.2. Appendages such as passive bilge keels or T-foils are common solutions for vessels on exposed routes. Independent research [3] shows that passive appendages have little influence on the motions, shown in Fig.8. The issue is compounded when considering the CTV has a low LD ratio hull form, unlike the higher LD hull form fast ferry types. These low LD ratio hull forms are characterised by a much stiffer WPA inertia and consequently requires a much larger restoring force to overcome the high inertial WPA stiffness.

The use of active T-foils has become a common feature on fast ferries where these systems have been shown to improve motions and reduce accelerations. However these are on hull forms with a much higher LD ratio where the restoring forces required pro-rata are much less than on low LD hull forms like CTVs. This results in increasing the size of the T-foils which means heavier equipment but more importantly an increase in drag, which slows the vessel down.

So the use of passive appendages will not provide sufficient effect to counter the higher motions that shall be experienced on these exposed routes for these short and heavy catamaran hull forms used by CTVs. Active appendages will not have much influence on the low LD hull forms but are better than passive. However, just not enough for such exposed routes owing to the forces required to compensate, renders such systems, like the gangways, large and heavy.

The evidence from the simple review is clear that a standard catamaran type hull form is not ideally suited for the higher sea states and a change in the hull form is required. The vessel may well be able to perform from a purely technical point of view, in terms of speed and payload capacity, but it is hindered by its own geometric limitations that influence its own natural motions and/or appendages damping them, and thus, renders operations in higher sea states uncomfortable and hazardous for transfers onto the towers.

What is also clear from the evidence of using heave compensated gangways on different vessel types, is that it points to the most obvious solution. The heave compensated gangways used on Swath vessels have been small compact and very successful. The required reduction in vertical displacement may be similar to that of a conventional vessel, in a similar sea state, but the reduction in forces and thus accelerations makes controlling such gangways significantly easier. Therefore if a heave compensated gangway works easily owing to the benign long period motions of a Swath, why not make the whole vessel "heave compensated" as such? With a hull form that has a roll period of 9 seconds, as MCS Swath 1 does, this becomes the "gangway". In other words the heave compensated gangway is the hull itself. A Swath hull form is therefore the most obvious hull form selection to minimise the accelerations and thus minimise the crew and technicians fatigue during long exposed transits and more importantly when transferring technicians on the tower.

In addition to the longer periods of motion, the stiffness of the waterplane area is much lower than a conventional CTV, which means the ability to provide damping via passive control surfaces is greatly improved. Small appendages such as bilge keels or fixed fins, can provide a significant reduction in amplitudes, which is not possible on conventional hull forms.

3.2 Powering Requirements

There are two arguments, myths even, with regards to Swaths. One is that of its resistance being a very draggy hull form and the power requirements merely to achieve a medium speed are cost prohibitive. The second is that they are so weight sensitive they are unable to carry high payloads. There is some truth in this, however, a CTV has a low LD ratio, in other words, it is draggy for its length. Bearing in mind that the existing CTVs at the time of design are generally built to remain below the 24m Laodline rules, we can use this as a basis for comparison. Using three typical hull forms: i) A typ catamaran CTV hull, ii) a generic high speed Swath hull and iii) a generic slow speed Swath hull and plotted on a speed versus resistance graph Fig.9, the results are rather illuminating. Because the typical CTV catamaran hull has such a low LD ratio is has a slightly higher resistance than compared to a high speed Swath hull form. The resistance curve of a typical slow speed Swath is clearly very draggy and is where most myths stem from, with regards to power requirements.

Given the sea conditions in Fig.6 and knowing their periods referring this to the encounter v speed chart in Fig.7, above 4.0seconds would be a good start at the lower end of the speed range and 2.5 seconds at the higher speed end. What is also obvious from the chart and requirements for a Swath hull form, Fig.7 indicates the conflicting requirements that must be equally satisfied; that of slow speed encounter periods and high speed encounter periods. A typical slow speed Swath hull form is optimised using appendages and the hull form itself to create large amounts of damping and natural periods sufficiently far removed from the expected wave conditions. The amount of damping on a Swath can reduce the vertical motions to almost zero. However, as the appendages are added to increase the damping, it increases the resistance considerably. This is shown by the slow speed Swath curve in Fig.9. Such a Swath is optimised for slow speed and stationary conditions to reduce the heave amplitude to near zero. This is to ensure the natural periods of motion are sufficiently far away from the stationary wave encounter periods. As can be seen in Fig.7 when a typical 30m CTV vessel is running at its service speed the period of encounter very closely matches the natural periods of the vessel. However, this encounter period changes from that of circa 2.5 seconds at 30 knots to 6.5 seconds when stationary. Yet typical waves of Hs=2.0m in the North Sea have wave periods of circa 6.5 seconds. This means the Swath must have its natural periods of motion sufficiently far away from 6.5 seconds to avoid any resonance – unfortunately this coincides with the natural periods of the high speed Swath hull form. This is where the conflicting requirements of slow speed/stationary conflict with that of high speed; how to make a Swath hull that has its natural periods of motion suitable for high speed - to reduce drag - and also at slow speed -without causing drag.

3.3 Optimising Appendages

The solution lies in the understanding of what a low WPA actually means. A low WPA is a low restoring force which means it is easy to control with external influences. A sensitive Swath is very easily recognised merely by walking across the deck, the slight changes in trim and list can be readily felt. Ad Hoc Marine Designs were responsible for the production design naval architecture of the Lockheed Martin SLICE©, and with just one or two persons walking across the deck a very noticeable list occurred. This is where the second myth about Swaths being unable to carry large payloads comes from. A Swath hull form is designed to be sensitive; it is its raison d'etre. This is mistaken for not being able to carry loads. A Swath is a fixed draft vessel, and the payload is designed into the hull form from the outset. Its 'sensitively' to minor perturbations, is irrelevant in this context, as it is sensitivity to changes in the arrangement of weights alone; since the fixed trim means fixed LCG.

Thus, being very sensitive to changes in weight distribution also has the same effect with a foil or fin that creates a lift force; it can easily restore the Swath's running attitude back to zero. If a Swath is tuned to be radically different from ocean swell encounter periods, such as the SSP Kaimalino Swath or the larger Kaiyo, it is extremely sensitive and when under way this sensitively results in a low longitudinal pitch stiffness. This gives rise to the commonly known Munk effeçt. These larger slow speed Swaths that have been built, make use of passive or active motion control fins both fwd and aft to ensure that any onset of pitch instability is corrected by the force from the fins provided by the passive fins or active motion control system.

What if a correctly designed motion control system could be used help to bridge the gap between slow speed requirements and that of high speed requirements. The high speed is obviously much easier, but if the fins could be larger to provide -additional control and damping than would otherwise be - and with a degree of control when stationary too - that would possibly solve the dilemma. The flaw in this argument is that fins which are lifting surfaces only work with forward speed to provide the lift. However, a motion control system that is able to provide heave compensation by actively moving the fins when stationary will have a significant effect on a sensitive Swath hull form. Couple this to additional appendages in the form of bilge keels and adding more damping to the hull shape, the combination will provide the slow speed motions that are desired.

3.4 Motion Control System

The active stabilization system installed on the Ad Hoc Designs 26 meter CTV, "MCS SWATH 1", is the current implementation and integration of Swath active control approaches and methods first developed and documented as part of the USN Swath development efforts [4, 5]. The system originally developed and implemented by Dr.Higdon has been extended, through further development and trials over the years, to include the contouring control objective in addition to the platforming mode. The use of unreliable depth pressure sensors to determine bow immersion, as needed to support earlier implementations of the contouring control mode, has been rendered obsolete by the availability of lower-cost but highly accurate microwave radar bow height sensors. The general arrangement of a modern Swath stabilization system, and as installed on the "MCS SWATH 1" CTV is shown in Fig.10.

Underway trials of MCS SWATH 1, conducted in seas exceeding 2 meters significant wave height, demonstrated that the performance of the active stabilization system was extremely good and as expected for this vessel and control system combination. In addition to reducing RMS pitch and roll motions to very low values, the system maintained vessel attitude – trim and list – to within vary narrow limits at all times from speeds of approximately 8 knots through full ahead, Fig.11.

What is new or novel about the stabilization system installed on the "MCS SWATH 1" CTV is the addition of a new control mode, developed for use when the vessel is lying to or at slow speeds ahead up to about 3 knots. This low- and zero-speed mode has, as its control objective, the reduction of bow heave motions; motion primarily results from pitching. The benefits of this control mode are twofold; the bow vertical motions are reduced during an approach to the landing on the turbine pylon, and the passenger comfort on board is enhanced during periods when the vessel is loitering on station between assigned transfer tasks.

Unlike the control of motions and vessel attitude underway using proportional control of the fin angles to vary fin lift forces, zero-speed control is effected by the use of impulse functions. The fins are "triggered" to rapidly move between their limit positions in response to a fuzzy-logic control algorithm that times the impulses to coincide with a specific magnitude and phase angle of a pending bow height deviation or excursion. The forces that create the impulse, the integral of force over the time duration of the fin sweep, result from the viscous resistance and change in momentum of the water that the fin creates as it flips from one limit angle to the other. Because the fin has more than 75% of its total planform area to the trailing side of the fin shaft, there is a net force and impulse created, as illustrated in Fig.12.

Although the resulting impulse values are not particularly large, their effect, properly timed and applied, on the very lightly damped but also lightly forced Swath pitch motions is quite significant when in wave conditions where the resonant pitch and bow heave behavior is being excited.

In Fig.13, are results from initial trials with MCS SWATH 1 Zero-Speed Heave (ZSh) Stabilization, off Denmark coast, 17 Sept 2016. The data from two contiguous sets of measurements is presented, showing the reduction in pitch rate, and hence bow heave, resulting from the toggling of fin positions in response to the damping control algorithm.

Sea conditions and the effects of engaging the zero-speed mode are as follows:

Loitering in 1.2-1.4m Hs seas, predominantly from stern

>58% reduction in RMS pitch rate
>70% reduction in RMS pitch angle
>70% reduction in RMS bow vertical displacement

Chart units: pitch angle (deg), pitch rate (deg/s), fin angles (degrees) Note: Fins are actually toggling between plus/minus 40 degrees. The Y-axis range is trimmed for clarity.

The unique zero-speed mode is fully effective with the low stiffness of a Swath hull form. The reduction in angles and amplitude of over 70% in moderate sea condition is evidence of the effectiveness of the system coupled to a low stiffness hull form. This achieves the objective of the design brief of a system that can bridge the gap between a high speed Swath hull form with that of a slow speed Swath hull form enabling the new hull form to maintain a high speed whilst augmenting the requirements at slow and stationary speeds.

4. IN-SERVICE OPERTATIONAL EXPERIENCE

The following account is based upon the experience and testimony of the Capt of the Swath vessel MCS SWATH 1, operated by Maritime Craft Services (Clyde) Ltd (MCS). MCS has been operating the vessel since September 2016 at various locations in the North Sea.

4.1 Motions

The first thing to notice on MCS SWATH 1 is the lack of motion. It is absolutely astonishing when the vessel can maintain its full crusing speed in seas exceeding Hs= 2.0m in any direction without any concerns for the safety of the vessel or the comfort of the passengers / technicians. What this means is when operating the vessel during transit the technician are free to move around the vessel can relax and enjoy hot refreshments without fear of spilling hot coffee on themselves. The technicians now arrive at their place of work relaxed and ready to start.

4.2 Pushing onto the Tower

The next benefit of the Swath hull form comes in to play on site. Once the procedure of pushing onto the tower begins there is no movement at the bow whatsoever, even up to and beyond Hs=2.0m. This clearly demonstrates a much safer and controlled environment for allowing the safe transfer to and from the tower. When assessing the safety of the transfer it is standard practice to gauge the bow movement by counting the number of rungs of the ladder the bow moves vertically. The most number of rungs, that is common for any conventional catamaran we have in our fleet, is 8 rungs in a sea of Hs=1.7-1.8m. However, with MCS SWATH1 the most the vessel has ever slipped was half a rung; and this only occurred once in exceptional circumstances and after the vessel slipped she held fast and allowed for a safe transfer to the tower.

4.3 Weight Sensitivity

With the nature of the Swath there is a critical balance that has to be considered at the early stages of the design. The usual thing you need to consider: service speed, cargo, fuel etc. are all much more critical on a Swath. What this means in practice is that in operations we are somewhat limited in the versatility of the vessel in respect to carrying load; as previously noted it is sensitive to changes in the LCG. With the wide variety of tasks required in building a windfarm the WFSV are required to carry out a wide range of operations from carrying not only technicians but to carrying heavy cargo and expecting to stay on site for up a week and longer if possible. This is very difficult to judge at the early stages of the design as each requirement will have an effect on something else, i.e. if it suddenly required to carry more cargo than the vessel is designed to carry, the vessel is no longer on the fixed draft marks thus the buoyancy (size of tubes) must increase and this increases drag and so effects the speed. This is not so evident on a conventional catamaran as when taking on additional payload load a conventional hull will merely sit on a deeper draft and lose some speed but as soon as the additional cargo is offloaded the speed will return. This is not so on a Swath as it is a fixed draft vessel and the displacement is fixed and designed into the vessel early on in the design process.

What this means as an operator we have to decide our requirements for the vessel early on, making compromises with range verses load etc. early on and working with what we have by replacing the fuel load with cargo load when required to carry cargo. It is all a bit of a balancing act and just a new skill to learn compared to the conventional hull form. But it does not hinder our operations.

4.4 Overall Impression

On the whole the experience of taking on MCS SWATH 1 has surpassed all our expectations of what a vessel of this size can achieve. The most obvious change noticeable from a seaman's perspective and provides much entertainment is when the vessel is steaming in rough seas and the bow hits a large wave. A seaman's natural instinct it to brace ready for the impact and the resulting slam and high accelerations, yet, nothing happens, it is truly amazing! Having been the skipper of many different types of CTVs, I honestly say that MCS SWATH1 is the best crew transfer vessel in Europe. I certainly wouldn't want to swap here for anything else I have seen out there.

5. CONCLUSIONS

The Swath hull form has been around since the early 1940s and is still considered by many as a concept despite many successfully operating modern examples. A Swath is also misunderstood in terms of what the hull form can achieve when compared to conventional vessels, such as catamarans.

Given the changing requirement of the CTV from inshore round 1 to the outer reaches of round 2 and the new round 3 windfarms, it is clear the design template - that of the existing small to medium heavy catamaran - is unsuitable for the sea conditions that dominate. Seakeeping that was previously a "nice to have" has been elevated to a "must have" requirement. Heave compensated gangways on existing catamaran CTVs are cumbersome and take up too much valuable foredeck space.

The ability to tune the natural frequencies of a Swath hull form away from the dominant wave periods a CTV shall encounter, makes the Swath hull form the obvious choice when faced with Hs=2.0-3.0m.

Manipulation of the Swath hull form to allow transfers when stationary, yielding low motions, and that of high speed transit, with minimal drag, is realised with advanced motion control systems that bridge the gap of the competing requirements. Thus making the whole vessel "motion compensated" and results in a vessel that has extremely low levels of acceleration when encountering Hs=2.0+m seas during transit and also when stationary during transfers in such sea states. This leads to the crew and technicians arriving safely at their destination, fresh and ready to work at the same time as increasing the operational window far beyond what is currently possible with conventional hull forms.

6. ACKNOWLEDGEMENTS

The authors would like to thank MCS and Island Engineering for providing the data of the vessel when the vessel has been running in-service and its experience gained so far.

7. REFERENCES

1. Warren NF, Rudgely G, "Designing a Special Purpose Swath to the HSC Code", The Safety of High Speed Craft, RINA Conf. Feb 1997, London.

2. OWA Offshore Foundation, "*Analysis of 10 years metocean model data*", October 2009.

3. Warren, N, Kecsmar, J, "Practical Design Aspects in the Hydrodynamics of High Speed Craft", R.I.N.A. Hydrodynamics Conference 24-25 November 1999

4. Fein, James A., "Seakeeping and Motion Control Trials of SSP Kaimalino in Sea States 4 and 5". DTNSRDC Report 81/015,.

5. Higdon,D.T. "Active Motions Reduction on SSP Kaimalino in a Seaway", Naval Ocean Systems Center.

6. Zarnick, Ernest E., "Vertical Plane and Roll Motion Stabilization of SWATH Ships", DTNSRDC Report 1199-01

8. AUTHORS BIOGRAPHY

John Kecsmar, co-founded Ad Hoc Marine Designs Ltd in 2005, and has been designing high-speed aluminium patrol boats, fast-ferries, SWATHs, and crew boats for almost 30 years. He has authored many technical papers on hydrodynamics and structural design of high speed vessels - focusing on fatigue of aluminium. He was previously Chairman of the Royal Institution of Naval Architects High Speed Craft committee (2012-16) and is currently a member of Lloyd's Register technical committee. He has just completed writing the draft for his book on the structural design and fabrication of aluminium high speed vessels at the same time as studying for his PhD.

J. William McFann - BSME, 1985 - Purdue University, Lafayette, IN. Member American Society of Naval

Engineers, Society of Naval Architects and Marine Engineers, and International Hydrofoil Society. Mr. McFann is currently the President of Island Engineering, Inc. (IEI), a company established in 1999 for the purpose of the development and testing of advanced marine vehicle designs and dynamic motion and flight control systems. From 1987 through 1999, Mr. McFann held positions as Design Engineer, Engineering Manager, and Director of Special Projects at Maritime Dynamics, Inc. (MDI), Lexington Park, MD. Mr. McFann managed key elements of the design, development and testing of numerous mono hull, catamaran, SWATH and SES craft.

David Low Master 200T, **g**raduated with a general degree in Yacht and Power Craft Design from Southampton institute. After a period of several years as a small boat builder and designer in 2008 become a Superintendent for Maritime Craft Services being responsible for the maintenance and day to day running of an extensive fleet of work boats, including 8 Damen 26m twinaxe vessels. In 2014 he became a Master on windfarm service vessels for Maritime Craft Services and was awared his Masters 200T certificate in 2015. After expressing a high degree of interested in Swath vessels he was appointed to MCS Swath1 as one of the senior skippers.

Figures:

Fig.1 Typical Transfer

Fig.2 Heave Compensated Gangway

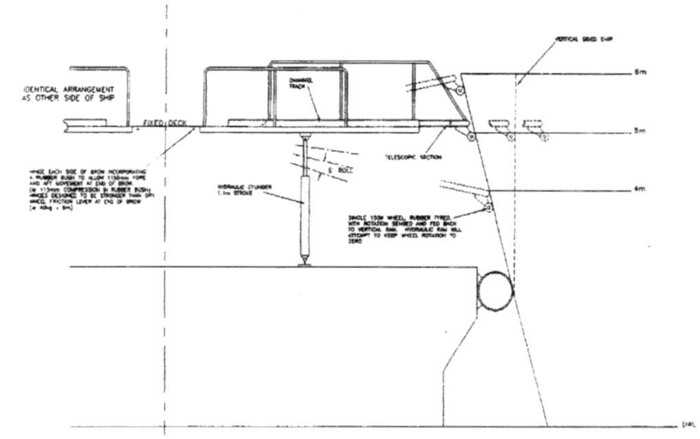

Fig.3 Transfer Brow on Swath

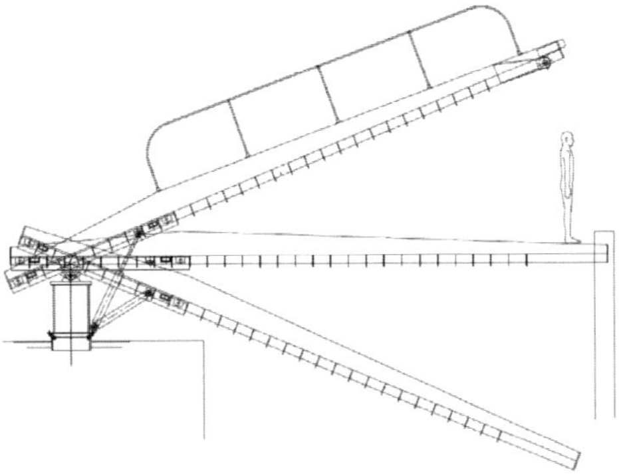

Fig.4 Foredeck Brow

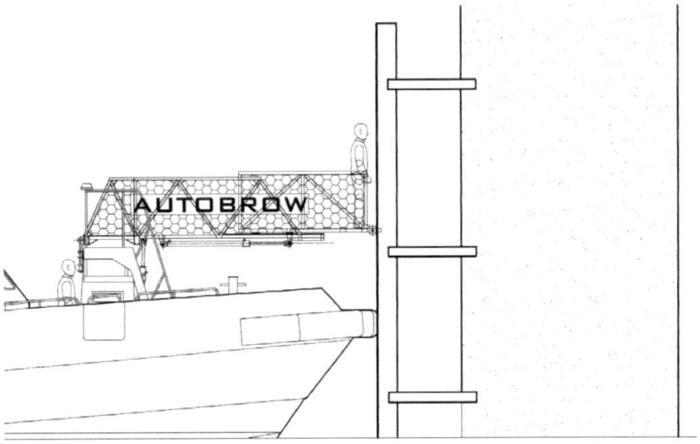

Fig.5 Autobrow on a 20m Catamaran

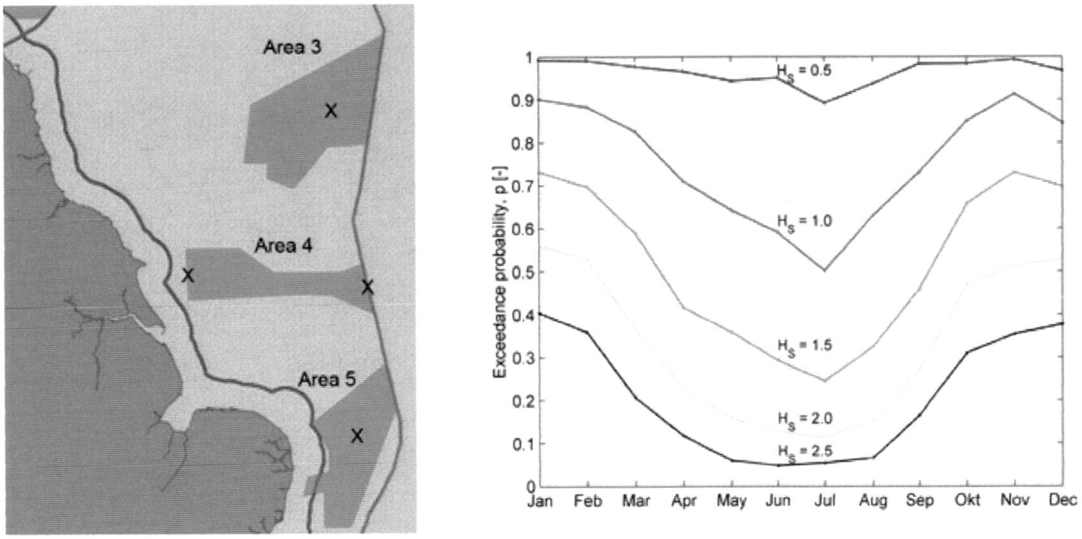

Data Extraction Points

Hs at Location 3

Fig.6 North Sea Wave Data

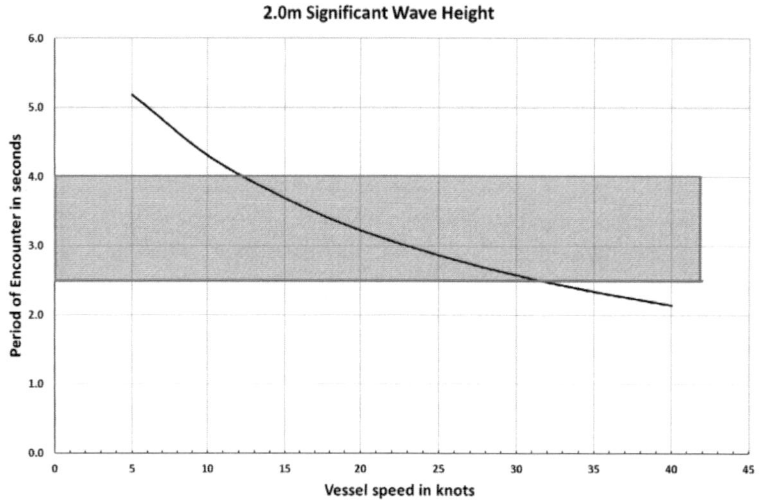

Fig.7 Periods of Encounter v Speed

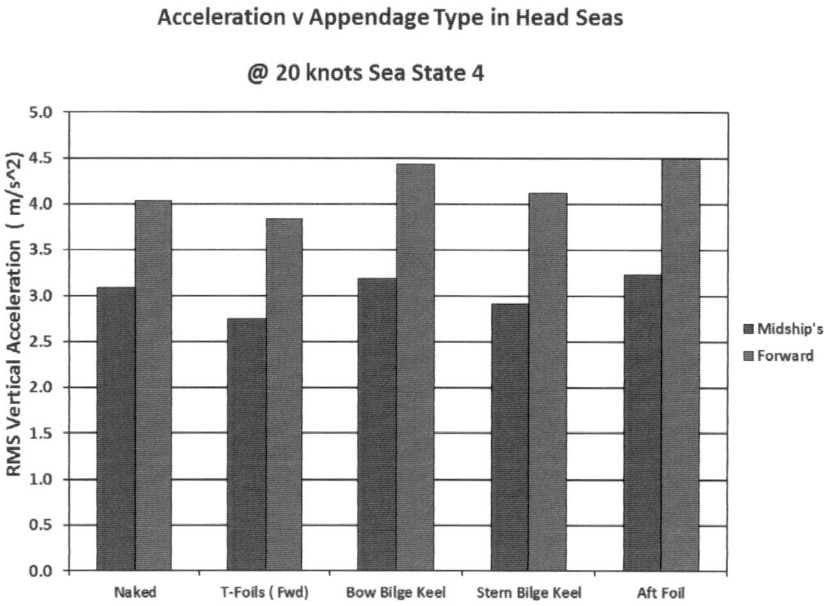

Fig.8 Effect of Appendages on Conventional Multihulls

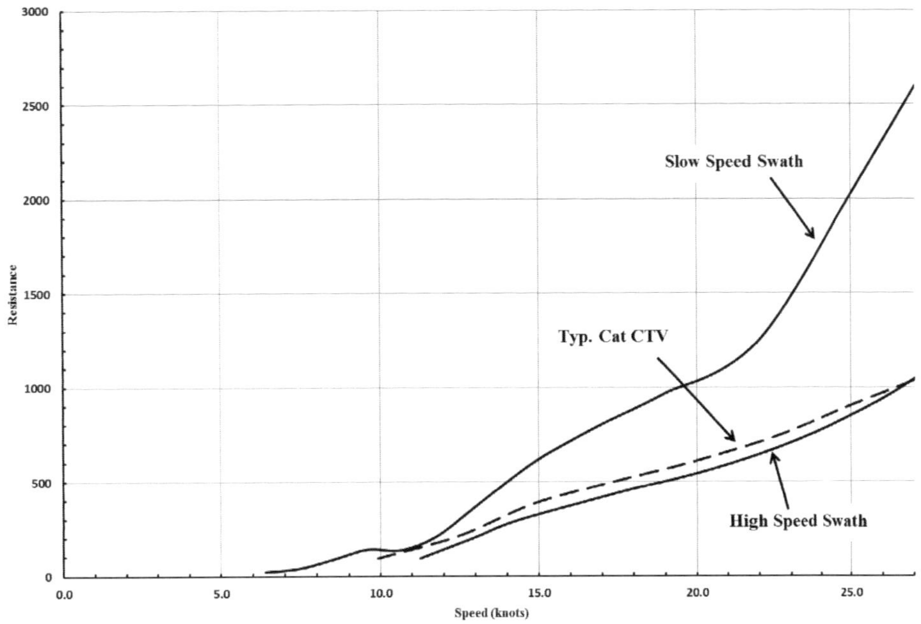

Fig.9 Resistance of Various Hull Form on an LWL of 23.8m

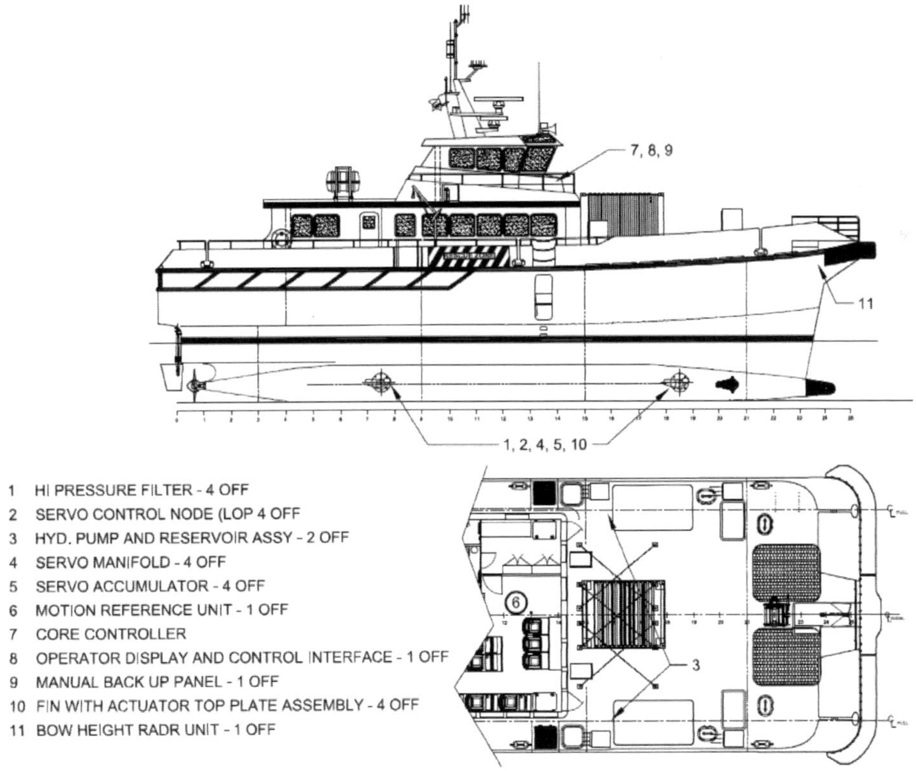

Fig.10 Arrangement of Motion control Sensors and Fin Assembly

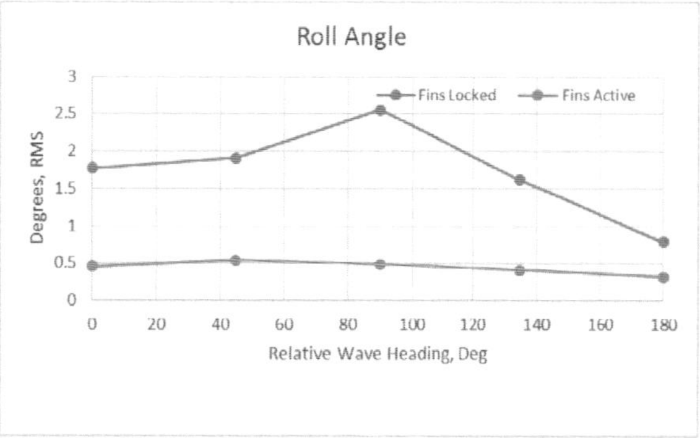

Roll Angles – Locked and Active Fins

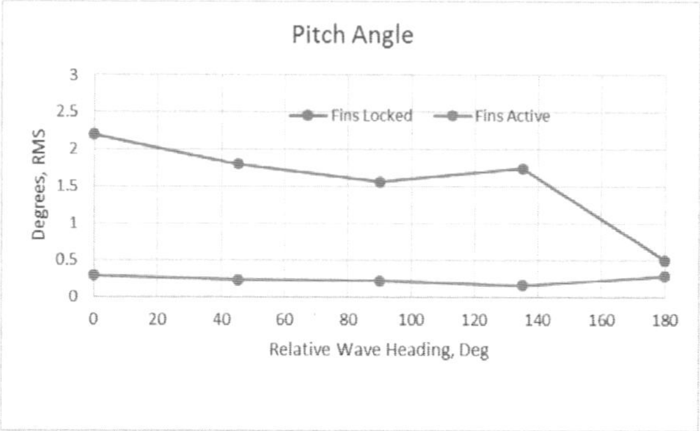

Pitch Angles – Locked and Active Fins

Fig.11 Roll and Pitch Angles of Fins

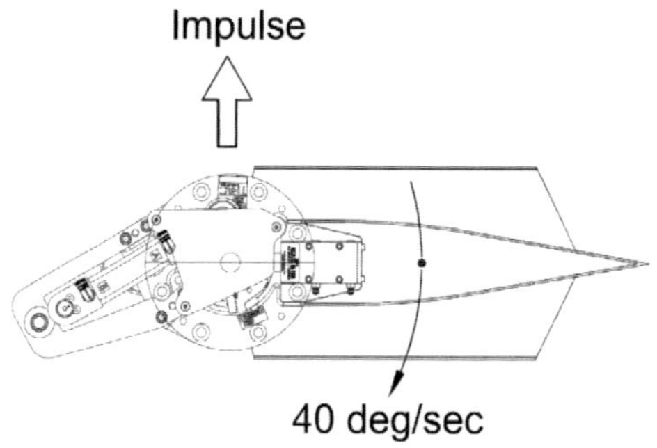

Fig.12 Fin Impulse Forces

ZS_h OFF

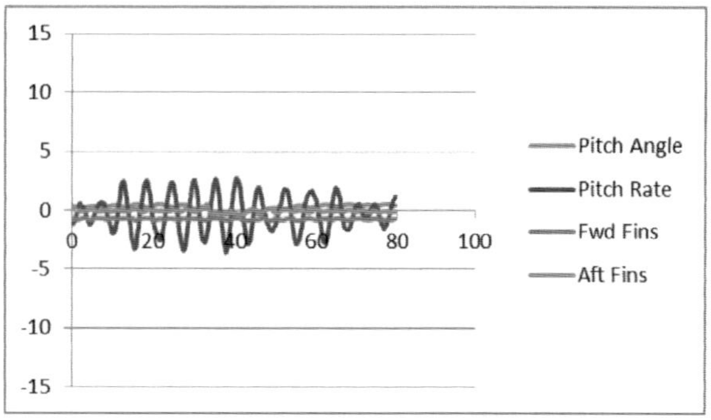

ZS_h ON

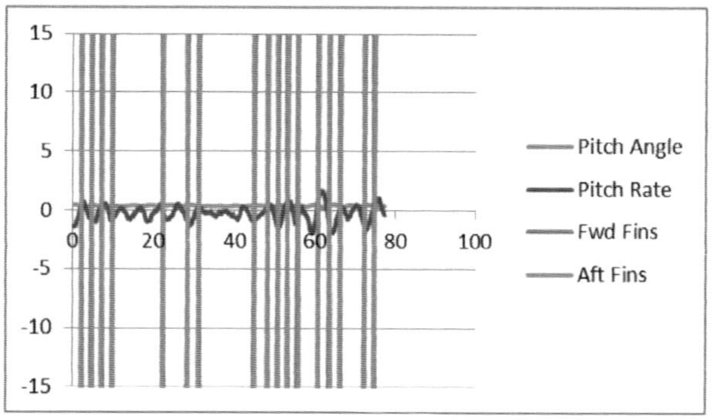

Fig.13 Effect of Zero-Speed Mode

SHIP SYNTHESIS MODEL FOR THE CONCEPTUAL DESIGN OF A SWATH WIND FARM SUPPORT VESSEL

O. V. Bondarenko, Admiral Makarov National University of Shipbuilding, UKRAINE

SUMMARY

Ship synthesis model for the conceptual design of a SWATH Wind Farm Support Vessel is developed. This model consists of a SWATH mathematical model and operational model. SWATH mathematical model is used for calculation of the main particulars. It includes the following main calculation units: ship main dimensions, resistance and power of the main engines, ship weight and weight subdivision, trim, stability, seakeeping, building cost.
Calculation of the economic efficiency indicators is performed in the ship operational model. The search for optimal values of the main particulars of the SWATH ships is carried out by solving the optimization problem using the genetic algorithm. As a goal function, the minimal cost of repairs and maintenance of the wind farm is used.
Ship synthesis model can be used to support the design characteristics of SWATH Wind Farm Support Vessels.

NOMENCLATURE

∇	Volume displacement (m^3)
∇_H	One lower hull displacement volume (m^3)
∇_S	Strut submerged volume (m^3)
β	Scale factor;
Γ	Gamma-function
δV_i	Coefficient which takes into account the loss of speed in a rough sea
Δ	Full load displacement (t)
η_D	Propeller or waterjet efficiency
η_G	Efficiency of the gear box
η_P	Payload coefficient
η_S	Shaft efficiency
θ	Integration variable
ρ	Form factor
ρ_W	Density of sea water (t/m^3)
$\sum_i R_{Wi}$	Individual wave resistance of every body that is a part of the SWATH ship (kN)
$\sum_i \sum_j \Delta R_{Wij}$	Additional wave resistance as a result of the wave systems interference (kN)
$2b$	Distance between center line of the hull (m)
A_{MH}	Midship area of the lower hull (m^2)
A_{r_i}	Required room area (m^2)
A_{WPS}	Waterplane area (m^2)
b_H	Ratio of the lower hull beam to its depth
B_{Box}	Box beam (m)
B_H	Lower hull beam (m)
B_S	Distance between lower hull centre line (m)
c_j	Unit cost of the material of the j-th weight item (GBP/t)
C_0	Coefficient that takes into account the weight of additional components (painting, welding material and margin
C	Vector of the parameters forming the design task
C_{adm}	Administration cost (GBP)
C_{BH}	Block coefficient of the lower hull
C_D	Cost of the main engines (GBP)
C_{dp}	Depreciation costs (GBP)
C_{eq}	Equipment cost (GBP)
C_{est}	Structure material costs and labour cost (GBP)
$C_{fuel}(i)$	Cost of fuel and lubricating oils (GBP)
C_{GB}	Cost of reduction gears (GBP)
C_i	Coefficient that takes into account the weight of other than the plate
C_{ins}	Insurance cost (GBP)
C_L	Total cost of extra labor required for the repair (GBP)
C_{ma}	Machinery cost (GBP)
C_{MH}	Midship section coefficient of the lower hull
C_O	Cost of electricity lost due to the turbine downtime (GBP)
C_P	Cost of spare parts and other materials necessary for the turbine repair or maintenance (GBP)
C_{PH}	Lower hull prismatic coefficient
C_{PR}	Cost of waterjets or screws (GBP)
C_{rm}	Repairing and maintenance costs (GBP)
C_{sal}	Crew costs (GBP)
C_S	Cost of service (personnel transfer) (GBP)
C_{TA}	Building cost (GBP)

C_{WL_A}	Aft waterplane area coefficient of the strut
C_{WL_F}	Fore waterplane area coefficient of the strut
C_{WPS}	Waterplane area strut coefficient
d	Ship draft (m)
D	Depth up to the main deck (m)
$D_F(i)$	Time of the turbine failure (hours)
D_H	Lower hull diameter (m)
DW	Deadweight (t)
$\overline{EPR}$	Average energy purchase rate, within duration of failure (GPB/kWh)
F	Objective function
$g_j(X)$	Ship performance requirements
g_{Str}	Superstructure volume density (t/m^3)
h_c	Ratio of the distance between lower hull center-line to the length of the ship
h_S	Strut depth (m)
$h_{3\%}$	Wave height of 3% probability (m)
H	Unit cost of the man-hours
H_{DB}	Depth of cross structure box (m)
H_H	Lower hull depth (m)
H_{DK}	Vertical clearance (m)
H_{SS}	Strut submerged depth (m)
k	Total number of the optimization problem's constraints
k_0	Wave number (1/m)
K	Number of weight items
k_W	Relative waterplane area
l_d	Ratio of the ship draft to the lower hull diameter
l_H	Relative length of lower hull
l_{NH}	Hull nose length to lower hull ratio
l_{NS}	Strut nose length to strut length ratio
l_S	Slenderness coefficient of strut
L_{Box}	Box length (m)
L_H	Lower hull length (m)
L_{OA}	Length overall (m)
L_S	Strut length (m)
L_{Work}	Number of man-hours spent (man-hours)
m	Number of the C vector parameters
M	Unit cost of structural material
MHR	Man-hour rate
M_s	Sea margin power
n	Number of independent variables;
n_a	Factor of the lower hull tail shape
n_f	Factor of the lower hull nose shape
n_h	Factor of the lower hull cross section shape
n_s	Strut nose and tail shape factor
N_C	Required number of repair technicians
p	Number of the optimization problem's constraints in the form of inequations
P_S	Engine power (kW)
$\overline{P}_{out}$	Average output wind power of turbine within duration of failure (kWh)
q_i	Density of the material hull part (kg/m^3)
R_{AA}	Aerodynamic resistance (kN)
R_{AP}	Appendages resistance (kN)
R_{AW_i}	Additional wave resistance for the *i*-th time step
R_F	Frictional resistance (kN)
R^n	Real-valued function in *n* variables
R_{SP}	Spray resistance (kN)
R_T	Total resistance (kN)
R_{T_i}	Ship resistance in calm water (kN)
R_W	Wave resistance (kN)
s_b	Strut setback
S_b	Strut setback (m)
S_i	Area of the *i* hull part (m^2)
t_i	Thickness of hull part (m)
t_S	Strut thickness (m)
T_{S_i}	Personnel delivery time (hour)
U	Ship speed (m/s)
V_0	Ship speed in calm water (kn)
V_{a_i}	Ship speed in a rough sea (kn)
V_{Sup}	Superstructure volume (m^3)
W_{BW}	Black water weight (t)
W_{CRG}	Deck cargo weight (t)
W_{CR}	Crew weight (t)
W_{FOil}	Fuel oil weight (t)
W_{FW}	Fresh water weight (t)
W_{GS}	General stores weight (t)
W_{Hull}	Hull weight (t)
W_{LS}	Lightship weight (t)
W_M	Machinery weight (t)
W_{Out}	Outfit weight (t)
W_{PL}	Maintenance technicians weight (t)
W_{si}	Plate weight of the main SWATH ship structural part
W_{SM}	Sea margin mass (t)
W_{Sup}	Superstructure weight (t)
W_{TAS}	Turbine Access System (TAS) weight (t)
W_{TE}	Weight of technician's equipment (t)

x_i^l	Lower bounds on the independent variable
x_i^u	Upper bounds on the independent variable
X	Vector of independent variables
Y	SWATH ship lower hull or strut offsets (m)

1. INTRODUCTION

Several types of vessels are currently used in the offshore wind energy for the personnel transportation. They are monohulls, catamarans, trimarans, small waterplane area twin hull ships (SWATH). One of the most promising types is the small waterplane area twin hull ships, particularly in view of one of the latest trends, i.e. removal of wind farms off the coast. This tendency leads to an increase in the time of transportation of the maintenance technicians to the wind farm turbines and an increase of comfort requirements – for the reduction of the fatigue of personnel.

2. MAIN BODY

In order to maintain offshore wind farms, 13 small waterplane area twin hull ships, seven TriSWATH ships and two semi-SWATH ships are currently used. The characteristics of these ships are shown in Table. 1.

Certain typical factors of SWATHs contribute to their application in transferring personnel to the offshore wind farms. They include:

- Excellent seaworthiness (large periods and small amplitude of motion).
- The course stability in rough sea;
- Absence of slamming and deck-flooding.
- The ability to develop high speed in rough sea.
- Large effective area of decks
- Less fatigue of the crew and personnel.

The design of SWATH ships is quite complicated. In the first place, it is due to the absence of experience in designing ships of this type. Today, a little more than 100 SWATH ships have been built and operated in the world. The SWATH ship itself, compared with catamarans and monohull ships, is a more complex subject and requires the designer to select more optimal parameters of the hull.

Hence, it seems rational to develop a synthesis model of the small waterplane area twin hull ships to transfer personnel on offshore wind farms. With the help of this model, it will be possible to select the optimal parameters of the hull and ship characteristics at the stage of conceptual design.

Table 1: Characteristics of SWATH Crew Transfer Ships

Vessel's name	CWind Astute	CWind Endurance	CWind Fulmar
Design OR Type		CTRUK SWATH20	CTRUK SWATH20
Delivery	2012	2014	2014
Hull Material	FRP	FRP	FRP
Transfer limit significant wave height, m	2	2	2
Length overall, m	19,9	20,0	20,0
Length bp, m	17,16		
Beam overall, m	7,5	7,8	7,8
Depth, m	3,55		
Design draft, m	1,55	1,49	1,49
Maximum draft, m		2,06	2,06
Design, deadweight, t	6		
Lightweight, t	25		
Design displacement, t	31	34	34
Full load displacement, t			
Main engines	Cummins QSM11	Cummins QSM 11	Cummins QSM 11
Total power, kW	2x 455	2x 455	2x 455
Max speed, kn	26	23	23
Service Speed, kn	20	19	19
Propulsion	FPP	WJ	WJ
Diameter propeller, m	0,9	–	–
Range, n.m.	1200	700	570
Deck area, m^2	56	56	56
Max deck cargo, t		2	2
Passengers	12	12	12
Crew	3	2–4	2–4

Vessel's name	Fob SWATH 1	Gesa	Sea Breeze
Design OR Type	S-Cat	S-Cat	
Delivery	2011	2012	2013
Hull Material	FRP	FRP	FRP
Transfer limit significant wave height, m	2,5		2,5
Length overall, m	27,0	29,5	24,76
Length bp, m	23,9	26,5	23,9
Beam overall, m	9,6	9,9	10,6
Depth, m	4,4	4,4	5,38
Design draft, m	1,75	1,5	1,8
Maximum draft, m	2,75	2,2	2,5
Design, deadweight, t			
Lightweight, t	82,5	93	90,5
Design displacement, t	–		
Full load displacement, t	125	153	138,7
Main engines	MAN D2862 LE422	MTU 10 V 2000 M72	MTU 10V 2000 M72
Total power, kW	2x749	2x900	2 x 900
Max speed, kn	22	23,4	23
Service Speed, kn	19	21	21
Propulsion	CPP	CPP	CPP
Diameter propeller, m	1,075		1,15
Range, n.m.		300	850
Deck area, m^2	30	55	50
Max deck cargo, t	10		10
Passengers	36	24	24
Crew	–	4	5

Vessel's name	Fob SWATH 7	Windea five	Sea Gale
Design OR Type			
Delivery	2016	2010	2013
Hull Material	FRP	Al	FRP
Transfer limit significant wave height, m		3,5	2,5
Length overall, m	33,19	26,4	24,76
Length bp, m	30,92	22,5	23,9
Beam overall, m	11	13,0	10,6
Depth, m	6	5,9	5,38
Design draft, m	1,5	2,7	1,8
Maximum draft, m	3,2		2,5
Design, deadweight, t		18	
Lightweight, t			90,5
Design displacement, t	–		
Full load displacement, t			138,7
Main engines		MTU 10V 2000 M72	MTU 10V 2000 M72
Total power, kW	2 x 1080	2 x 900	2 x 900
Max speed, kn	24	18	23
Service Speed, kn		16	21
Propulsion		FPP	CPP
Diameter propeller, m		1,3	1,15
Range, n.m.		500	850
Deck area, m^2			50
Max deck cargo, t		3,5	10
Passengers	40	12	24
Crew		3	5

Vessel's name	Sea Storm	Sea Hurricane	Lina 1
Design OR Type			
Delivery	2013	2013	2014
Hull Material	FRP	FRP	FRP
Transfer limit significant wave height, m	2,5	2,5	2,5
Length overall, m	24,76	24,76	24,76
Length bp, m	23,9	23,9	23,9
Beam overall, m	10,6	10,6	10,6
Depth, m	5,38	5,38	5,38
Design draft, m	1,8	1,8	1,8
Maximum draft, m	2,5	2,5	2,5
Design, deadweight, t			

Vessel's name	Sea Storm	Sea Hurricane	Lina 1
Lightweight, t	90,5	90,5	90,5
Design displacement, t			
Full load displacement, t	138,7	138,7	138,7
Main engines	MTU 10V 2000 M72	MTU 10V 2000 M72	MTU 10V 2000 M72
Total power, kW	2 x 900	2 x 900	2 x 900
Max speed, kn	23	23	23
Service Speed, kn	21	21	21
Propulsion	CPP	CPP	CPP
Diameter propeller, m	1,15	1,15	1,15
Range, n.m.	850	850	850
Deck area, m^2	50	50	50
Max deck cargo, t	10	10	10
Passengers	24	24	24
Crew	5	5	5

Vessel's name	MCS SWATH 1	Cable Bay	World Mistral
Design OR Type	Typhoon Class	Tri SWATH 27	TriSWATH
Delivery	2016	2013	2012
Hull Material	Al	Al	Al
Transfer limit significant wave height, m	3,5		2,5
Length overall, m	26,87	27,20	24,0
Length bp, m	23,5	23,0	
Beam overall, m	9,8	10,5	9,4
Depth, m	4,8	3,7	4,6
Design draft, m	2,1	2,1	2,2
Maximum draft, m			2,6
Design, deadweight, t		13,74	10,0
Design displacement, t	95		
Main engines	MAN: D 2862 LE 466	MTU 10V 2000 M72	4 x Scania DI13 070
Total power, kW	2 x 1029	2 x 900	4 x 368
Max speed, kn	24	25	25
Service Speed, kn	22	23	20
Propulsion	CPP	WJ	CPP
Diameter propeller, m		–	1,15
Range, n.m.	800	350	
Deck area, m^2	96	42	83
Max deck cargo, t	5	4,5	10
Passengers	12	12	12
Crew	3–4	3	2/4

The main requirements for the ships to operate offshore wind farms are as following:
- high speed which is important for fast delivery of service personnel from the base to the turbine requiring repair or maintenance;
- good seakeeping to transfer personnel in rough sea;
- presence of spaces on the deck to accommodate spare parts and equipment;
- a small cost of construction and operation;
- reliability and safety.

In order to implement these requirements, the designer needs to find many trade-offs. For example, the requirement of a small construction cost leads to selection of the minimum size of the ship and its displacement. On the other hand, the smaller the displacement, the more difficult it is to ensure a sufficiently high seakeeping (without using special devices). And seakeeping characteristics change with increasing displacement rather slowly: approximately proportional to the cube root of displacement [1].

In this case, it is rational to apply the parametric optimization in order to make better decisions.
Parametric optimization is used to find a set of parameters $X = \{x_1, x_2, ..., x_n\}$ (independent variables), giving the best situation, with respect to a chosen objective, for the considered system.

The optimization problem may in general be written as

$$\min_{X \in R^n} F(X, C)$$

$$\text{subject to a: } g_j(X) = 0, \quad j = p+1, ..., k,$$

$$g_j(X) \geq 0, \quad j = 1, ..., p,$$

$$x_i^l \leq x_i \leq x_i^u, \quad i = 1, ..., n$$

where $X \in R^n$ is the vector of independent variables, $F(X,C)$ is the objective function, $C(C_1,...,C_m)$ is a vector of the parameters forming the design task**;** m is a number of the C vector parameters; n is a number of independent variables; k is a total number of optimization problem constraints; p is a number of optimization problem constrains in the form of inequalities; R^n is n-dimensional Euclidean space; g is the set of constraints; x_i^l is lower bounds on the independent variable, x_i^u is upper bounds on the independent variable; x_i is values of the independent variables.

The vector C consists of the main characteristics of the design task vessels such as required service speed (kn), number of Maintenance Technicians, endurance (day), hull and superstructure material (aluminum Alloy , glass-reinforced plastic), number of struts (single, tandem), type of machinery (medium speed diesel, high speed diesel, diesel electric), type of propulsor (propeller. waterjet), weight of deck cargo, t.

The vector of independent variables are: x_1 – relative length of lower hull l_H; x_2 – slenderness coefficient of strut l_S; x_3 – waterplane area strut coefficient C_{WPS}; x_4 – relative waterplane area k_W; x_5 – ratio of the distance between lower hull center-line to the length of the ship; x_6 – ratio of the ship draft to the lower hull diameter l_d; x_7 – ratio of the lower hull beam to its depth b_H; x_8 – lower hull prismatic coefficient C_{PH}; x_9 – factor of the lower hull nose shape n_f; x_{10} – factor of the lower hull tail shape n_a; x_{11} – factor of the lower hull cross section shape n_h; x_{12} – strut nose and tail shape factor n_s; x_{13} – hull nose length to lower hull ratio l_{NH}; x_{14} – strut nose length to strut length ratio l_{NS}; x_{15} – payload coefficient η_P; x_{16} – strut setback s_b.

The system of optimization problem constraints includes variable bounds and one-way limits on vessel characteristics.

As the objective function of the optimization problem $F(X,C)$, the cost repair and maintenance of wind farm is used

$$F(X,C) = C_P + C_S + C_O + C_L,$$

where C_P – cost of spare parts and other materials necessary for the turbine repair or maintenance; C_S – cost of service (personnel transfer); C_O – cost of electricity lost due to the turbine downtime; C_L – total cost of extra labor required for the repair.

The cost of electricity lost due to the downtime of the i-th turbine is determined according to the work data [2] using the following formula:

$$C_O(i) = D_F(i)\overline{P}_{out}\overline{EPR},$$

where $D_F(i)$ – time of the turbine failure, hours; $\overline{P}_{out}$ – average output wind power of turbine within duration of failure, kWh; $\overline{EPR}$ – average energy purchase rate, within duration of failure, GPB/kWh;

The cost of works related to the fault repair is determined as follows

$$C_L(i) = D_F(i) \cdot N_C \cdot MHR,$$

where N_C – required number of repair technicians (Table 2); MHR – man-hour rate.

Table 2: Basic standards for the maintenance/repair of the wind turbine parts [3]

Failure Category	Failure Rate, times/ year	Repair Time, hours	Required Techni-cians	Repair Cost, GBP
Manual restart	8.79	1	2	–
Blades	1.48	8	2	20300
Pitch System	0.08	5	2	7150
Hub	0.185	4	2	4300
Main Shaft & Bearings	0.185	4	2	14000
Gearbox	0.08	4	2	3250
High speed shaft	0.19	3	2	3250
Mechanical Brake	0.04	2	2	2500
Generator	0.08	4	2	12000
Control system	0.24	4	2	4150
Yaw System	0.12	3	2	10800
Hydraulic Services	0.12	3	2	1300
Power Electrics	0.24	3	2	4150
Transformer	0.02	5	2	15400
Tower	0.19	8	2	5000

The value of the cost of service:

$$C_S = C_{dp} + C_{rm} + C_{sal} + C_{ins} + C_{adm} + \sum_{i=1}^{m} C_{fuel}(i),$$

where C_{dp} – depreciation costs; C_{rm} – repairing and maintenance costs; C_{sal} – crew costs; C_{ins} – insurance cost; C_{adm} – administration cost; $C_{fuel}(i)$ – cost of fuel and lubricating oils;

The search for optimal values of the independent variables is performed using a genetic algorithm [4]. In this case, the search process starts with the generation of the initial population, in other words – sets of values of independent variables (Figure 1). For each of this set, its characteristics, such as main dimensions, power of main engines, items of weight ship, stability and seakeeping parameters, are calculated with the help of a mathematical model of the ship. Then, using the operation model economic indicator of the ship is calculated. These values are used as the optimization criterion for estimating the effectiveness of the ship with the main dimensions (defined by a set of independent variables) and the formation of a new population according to one of the known selection rules.

The searching process is repeated until optimal values are received.

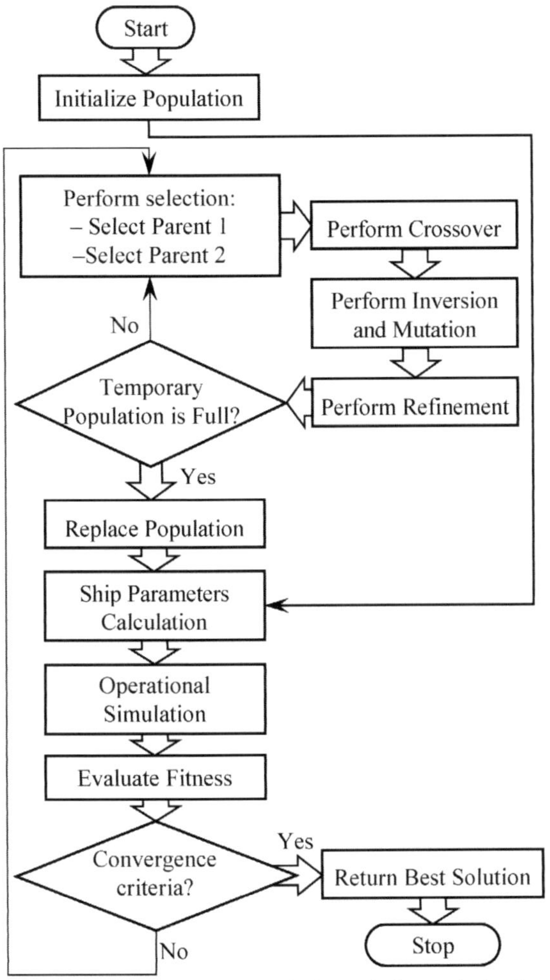

Figure 1: Block diagram solution of the optimization problem

Let us look at the calculation algorithms used in the synthesis model in more details. This model includes a mathematical model of SWATH ships and the operating model.

The mathematical (parametric) model of SWATH ships is used to calculate the characteristics of SWATH ships. The block diagram of the model is shown in Figure 2. This model includes various design relationships for calculating areas, volumes, sizes, weights, stability and costs of SWATH ships. These relationships are based on many technical literature sources.

For each set of independent variables, the calculation of SWATH ships characteristics is performed in the following sequence:

Waterplane area, m^2

$$A_{WPS} = \nabla^{2/3} k_W .$$

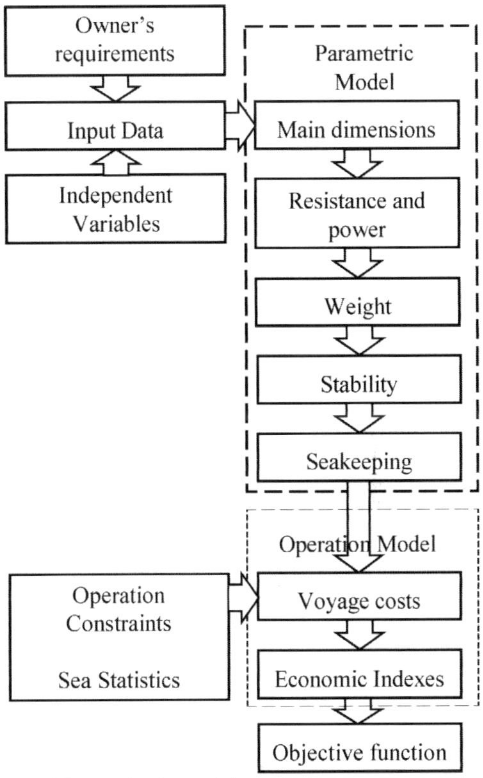

Figure 2: Ship synthesis model

Fore waterplane area coefficient of the strut

$$C_{WL\_F} = \frac{n_s}{1+n_s} .$$

Aft waterplane area coefficient of the strut

$$C_{WL\_A} = \frac{n_s}{1+n_s} .$$

Strut length, m

$$L_S = \sqrt{\frac{A_{WPS} l_S}{2C_{WPS}}} .$$

Strut thickness, m

$$t_S = l_S / L_S .$$

Midship section coefficient of the lower hull

$$C_{MH} = \frac{\sqrt{\pi}\cdot\Gamma\left(\frac{1}{n_h}\right)}{2^{\frac{2}{n_h}}\cdot n_h\cdot\Gamma\left(\frac{1}{2}+\frac{1}{n_h}\right)} .$$

Block coefficient of the lower hull

$$C_{BH} = C_{MH} C_{PH} .$$

Lower hull beam (m) from the equation solution

$$\frac{2C_{BH}l_H B_H^3}{b_H^{3/2}} + \frac{A_{WPS}(l_d - 1)B_H}{b_H} - \nabla = 0.$$

Lower hull depth, m

$$H_H = \frac{B_H}{b_H}.$$

Lower hull diameter, m

$$D_H = \sqrt{H_H B_H}.$$

Lower hull length, m

$$L_H = l_H D_H.$$

Ship draft, m

$$d = l_d H_H.$$

Midship area of the lower hull, m^2

$$A_{MH} = C_{MH} H_H B_H.$$

One lower hull displacement volume, m^3

$$\nabla_H = C_{PH} A_{MH} L_H.$$

Strut submerged depth, m

$$H_{SS} = d - H_H.$$

Strut submerged volume, m^3

$$\nabla_S = \frac{A_{WPS} H_{SS}}{2}.$$

Strut setback, m

$$S_b = s\_bL_H.$$

Length overall, m

$$L_{OA} = \max(L_S + S_b; L_H).$$

Box length, m

$$L_{Box} = \min(L_S + S_b; L_{OA}).$$

Vertical (box) clearance, m

$$H_{DK} = \max\left(0,75h_{3\%}; 0,625\sqrt{B_H H_H}\right).$$

Strut depth, m

$$h_S = H_{SS} + H_{DK}.$$

Distance between lower hull centre line, m

$$B_S = h_C L_H.$$

Box beam, m

$$B_{Box} = B_S + B_H.$$

Depth of cross structure box, m

$$H_{DB} = (B_{Box} - 2B_H)/7,5.$$

Depth up to the main deck, m

$$D = H_H + h_S + H_{DB}.$$

The power of the main engines can be calculated as follows:

$$P_S = \frac{R_T U}{\eta_D \eta_G \eta_S}(1 + M_s),$$

where R_T – total resistance, kN; U – ship speed, m/s; η_D – propeller or waterjet efficiency; η_G – efficiency of the gear box; η_S – shaft efficiency; M_S – coefficient of sea margin.

The propeller or waterjet efficiency η_D depends on the type of propulsion device. The algorithm given in the publication [5] is used for waterjets, and the one published in [6] is used for propeller.

For defining the total resistance R_T in calm water there was used the following formula:

$$R_T = R_F + R_W + R_{SP} + R_{AP} + R_{AA},$$

where R_F, R_{AP}, R_{AA} – frictional, appendages and aerodynamic resistance; R_W, R_{SP} – wave and spray resistance.

The calculation of frictional, appendages and aerodynamic resistance is performed by the known dependences of the ship theory considering the SWATH ship construction peculiarities [7, 8].

Spray resistance calculation is based on the results of the model tests, as described in [9].

The wave resistance of the small waterplane area twin hull ship is defined according to the formula, kN:

$$R_W = \sum_i R_{Wi} + \sum_i \sum_j \Delta R_{Wij},$$

where $\sum_i R_{Wi}$ – individual wave resistance of every body that is a part of the SWATH ship , kN; $\sum_i \sum_j \Delta R_{Wij}$ – additional wave resistance as a result of the wave systems interference, kN. The wave resistance is defined for those bodies such as underwater hull, fore and aft struts. There is also an opportunity to calculate the resistance for the Slice type ships and single-hull SWATH ship.

In order to calculate the components of the SWATH ship wave resistance there was used the Michell integral:

$$R_W = \frac{4\rho_w g^4}{\pi U^6} \int_0^{\pi/2} \left(I^2 + J^2\right) w^2 \sec^5 \theta \, d\theta,$$

where

$$J(\theta) = -\int_{-\frac{L}{2}}^{\frac{L}{2}} \int_{-T}^{0} Y(x,z) \exp\left(k_0 z \sec^2\theta\right) \cos\left(k_0 x \sec\theta\right) dx dz;$$

$$I(\theta) = \int_{-\frac{L}{2}}^{\frac{L}{2}} \int_{-T}^{0} Y(x,z) \exp\left(k_0 z \sec^2\theta\right) \sin\left(k_0 x \sec\theta\right) dx dz;$$

θ – integration variable; $k_0 = \frac{g}{U^2}$ – wave number, 1/m; $w = 2\cos\left(k_0 b \sec^2\theta \sin\theta\right)$; $2b$ – distance between center line of the hull, m; ρ_w –sea water density, t/m^3; U – ship speed, m/s; $Y(x,z)$ – SWATH ship lower hull or strut offsets, m.

Calculation of full displacement is performed according to the following dependence:

$$\Delta = W_{LS} + DW,$$

where Δ – full displacement, t; W_{LS} – lightship weight, t; DW – deadweight, t.

WFSV deadweight can be calculated by use of the following relation:

$$DW = W_{PL} + W_{FOil} + W_{CR} + W_{FW} + W_{GS} + \\ + W_{TE} + W_{BW} + W_{CRG} + W_{TAS}$$

where W_{PL} – maintenance technicians weight, t; W_{FOil} – fuel oil weight, t; W_{CR} – crew weight, t; W_{FW} – fresh water weight, t; W_{GS} – general stores weight, t; ; W_{TE} – weight of technician's equipment, t; W_{BW} – black water weight, t; W_{CRG} – deck cargo weight, t; W_{TAS} – Turbine Access System (TAS) weight, t.

The part of these components are the specified characteristics (deck cargo weight, TAS weight), and another part of them can be identified by use of the known relations of the ship design theory.

The algorithm for determination of items of the lightship displacement of WFSV is based on the data given in the articles [10, 11]. According to them, the lightship displacement W_{LS} is calculated as a sum of the following items: the hull weight W_{Hull}, the machinery weight W_M, the outfit weight W_{Out} and the sea margin weight W_{SM}:

$$W_{LS} = W_{Hull} + W_M + W_{Out} + W_{SM}.$$

According to [10] SWATH ship hull weight is estimated through the structural part thickness taking to account the operation load and requirements of the Ship-Classification Society:

$$W_{Hull} = (1 + C_0)\sum_i (1 + C_i) W_{si},$$

where $C_0 = 0,085$ – coefficient that takes into account the weight of additional components (painting, welding material and margin); C_i – coefficient that takes into account the weight of other than the plate (stiffeners); W_{si} – plate weight of the main SWATH ship structural part: lower hulls, struts, sponsons, box, inside decks and platforms, longitudinal and transverse bulkheads.

The plate weight of the SWATH ship hull structural parts is defined according to the following dependence

$$W_{si} = 0,001\sum_{i=1}^{n} S_i t_i q_i \text{, t,}$$

where S_i, t_i, q_i – area (m^2), thickness (m) and density of the material (kg/m^3) of the i hull part appropriately.

The surface area S_i of the structural parts of the SWATH hull is determined using the parametrical model and is directly connected with independent variables.

For the ships with glass-reinforced plastic hulls the similar coefficients are difficult to obtain. That's why according to the calculations for several small waterplane area twin hull ships made of glass-reinforced plastic there was obtained the following dependence between the hull weight and ship displacement:

$$W_{Hull} = 0,2168\Delta + 4,6129.$$

The superstructure weight at the first approximation is calculated depending on the material by the formula

$$W_{Str} = g_{Str} V_{Sup} \text{, t,}$$

where V_{Sup} – the superstructure volume with regard for the wheelhouse, m^3; g_{Str} – the superstructure volume density, t/m^3.

The other lightship weight groups with some changing and improvements are defined by formulas [11].

The estimation of building cost ship at the initial design stage is performed by the following relationships [12]:

$$C_{TA} = C_{est} + C_{eq} + C_{ma},$$

where C_{est} is the structure material costs and labour cost; C_{eq} is the equipment cost; C_{ma} is the machinery cost.

The value of C_{est} is determined by the following formula:

$$C_{est} = [1,1 W_{Hull} M] + [W_{Hull} L_{Work} H],$$

where M is the unit cost of structural material; H is the unit cost of the man-hours; L_{Work} is the number of man-hours spent, man-hours.

The machinery cost is

$$C_{ma} = 1,4(C_D + C_{GB} + C_{PR}),$$

where C_D is the cost of the main engines; C_{GB} is the cost of reduction gears; C_{PR} is the cost of waterjets or screws.

To calculate the equipment cost the following expression is used:

$$C_{eq} = \sum_{j=1}^{K} c_j W_j,$$

where W_j is the mass of the j-th weight item (t); c_j is the unit cost of the material of the j-th weight item; K is the number of weight items.

Further, the trimming parameters, the transverse and longitudinal stability, volume and others are calculated in the mathematical model.

The operation model describes the functioning of all operations of the SWATH ships associated with operating of offshore wind farms.

The well-known features are the following: the number of wind turbines in the park, the distance between the park and the base of supply vessels, the distance between the wind turbines, their locations scheme.

It is expected that the ships can perform voyages in the daytime when the wave height and wind speed do not exceed the maximum permissible value.

The departure of the personnel is undergone in the event of an accident or a failure of a wind turbine. The list of possible faults, the time needed to resolve them, the number of technicians required for the repair are set on the basis of the study of statistical data (Table. 2). It is assumed that the probability of failure / malfunction of turbines is the same for the entire wind farm.

The position of each turbine of the wind farm is set due to the appropriate coordinates. Then, the computer simulation of the wind farm is performed, during which the process of simulated emergency events and turbines failures are modelled. The flow of emergency events (failure rate) for wind turbines life cycle is represented as the Weibull law [3]:

$$h(t) = \rho\beta t^{\beta-1},$$

where β is the scale factor; ρ is the form factor.

The process of emergency or repair simulation works as follows. At any given time, the generator modelled a pseudorandom number R, which is set by a predetermined uniform distribution in the range from 0 to 1. The event of turbine failure is considered to be occurred when:

$$R > \left[1 - h(t)\right]\frac{\delta t}{8760}.$$

If it is the operating time and weather conditions allow to repair, ship course and repairing works of the turbine are modelled. During the process of modelling, the distance of ship passage $Dist_i$ is determined on the map. Next, we conduct the calculation of the following parameters:

- additional resistance in waves according to the current weather conditions R_{AW_i};
- loss of ship speed in a rough sea;
- personnel delivery time $T_{S_i} = \dfrac{Dist}{V_{a_i}}$;
- the variable components of the operating ship costs;
- coefficient of operational use of the ship.

When calculating the personnel delivery time we taken into account the current weather conditions in the area of wind farms and determine the ship speed in a rough sea:

$$V_{a_i} = V_0 - \delta V_i,$$

where V_0 is the ship speed in calm water; δV_i is the coefficient which takes into account the loss of speed in a rough sea:

$$\frac{\delta V_i}{V_0} = \sqrt{1 + \frac{R_{AW_i}}{R_{T_i}}},$$

where R_{AW_i} is the additional wave resistance for the *i*-th time step; R_{T_i} is the ship resistance in calm water.

If it is time off or weather conditions are unsuitable, then demurrage is modelled and the wind farm loss owning to a broken turbine downtime is defined.

The proposed mathematical model and operating model were implemented as a software package «SWATH CTV».

3. CONCLUSIONS

The developed synthesis model allows to determine the optimal characteristics of the small waterplane area twin hull ships to transfer personnel on offshore wind farms at the stage of conceptual design. Also, with this model it is possible to solve other issues, for example, to determine the effectiveness of the use of SWATH CTV for a specific wind farm.

4. REFERENCES

1. DUBROVSKY, V.A., 'Effective and inexpensive patrol ship with helicopters and unmanned vehicles', *Navigation and Marine Researches: selected reports of the Third Sakhalin regional marine scientific conference, Feb-15-16*, 2011.

2. KAHROBAEE, S., ASGARPOOR, S., 'Risk-Based Failure Mode and Effect Analysis for Wind Turbines (RB-FMEA)', *Proceedings of the North American Power Symposium (NAPS), Boston, USA*, 2011.

3. DALGIC, Y., 'Optimum CTV fleet selection for offshore wind farm O&M activities', *Safety and Reliability: Methodology and Applications, London*, 2014.

4. SIVANANDAM, S.N., DEEPA, S.N., 'Introduction to Genetic Algorithms', *Springer Publ.*, 2007.

5. VAN TERWISGA, J. C., 'A parametric propulsion prediction method for waterjet driven craft', *Proceedings of the 4th International Conference on Fast Sea Transportation (FAST '97), Sydney*, 1997.

6. OOSTERVELD, M., OOSAMEN, P., 'Further Computer Analyzed Data of the Wageningen B-Screw Series', *International Shipping Progress*, 1975.

7. NETHERCOTE, W.C., SCNMITKE, R.T., 'A Concept Exploration Model for SWATH Ships', *The Naval Architect Journal, vol. 124, № 5*, 1982.

8. BERTRAM, V., 'SWATH Ship Design Formulae Based on Artificial Neural Nets', *Journal of Ship Technology, vol. 3, № 1*, 2007.

9. CHAPMAN, R.B., 'Spray drag of surface piercing struts', *AIAA/SNAME Advanced Vehicles Conference, Annapolis, july*, 1972.

10. BONDARENKO, O.V., BOIKO, A.P., SEROPYAN, I.R., 'Determination of the main characteristics of the small waterplane area twin hull ships at the initial stage of design', *Polish Maritime Research, Vol. 20, Issue 1(77)*, 2013.

11. BONDARENKO, O. V., 'Preliminary weight estimates for wind farm support vessel', *Design & Construction of Wind Farm Support Vessels, 30-31 March 2016, London, UK*, 2016.

12. MORAES, H. B., VASCONCELLOS, J. M., ALMEIDA, P. M., 'Multiple criteria optimization applied to high speed catamaran preliminary design', *Ocean Engineering, 34*, 2007.

5. AUTHORS BIOGRAPHY

Oleksandr Bondarenko holds the current position of associate professor at Admiral Makarov National University of Shipbuilding (Nikolaev/Ukraine). He is a Fellow of the Royal Institution of Naval Architects. His key research area is WFSV, SWATH and multi-hull ship design, simulation modeling, optimization, genetic algorithms and their application in ship design, and the development of advanced ship design methodologies. His previous experience includes development of the statistical method of design based on the study and synthesis of the principal particulars of WFSV.
E-mail: abondarenko70@mail.ru

A SUMMARY OF THE ENGINEERING PROCESS AND THE MODELLING TOOLS USED IN THE DESIGN OF A NEXT GENERATION WFSV INCORPORATING A SUSPENSION BASED RIDE CONTROL SYSTEM PROVIDING IMPROVED TRANSIT AND TRANSFER PERFORMANCE

M J Longman, Nauti-Craft Pty Ltd, Western Australia

SUMMARY

A 24m (LWL) Wind Farm Service Vessel incorporating a revolutionary suspension based ride control system is being designed and built for operation in the North Sea. The vessel's primary design features include improved transit comfort in extreme sea conditions and safe technician transfer in seas with a 2.5m significant wave height. The proprietary ride control system decouples the hulls from the superstructure through the use of a passive reactive interlinked hydraulic system capable of up to 2.8m of suspension travel. The following paper provides a summary of the engineering design process including the relevance and effectiveness of the modelling tools in enabling the development of this unique vessel to a level that is acceptable to classification societies.

1. INTRODUCTION

This paper covers the engineering process and simulation techniques employed in the design of a 24m wind farm service vessel (WFSV) incorporating the Nauti-Craft interconnected hydraulic suspension and ride control system.

The catamaran-based WFSV is currently being designed by Southerly Designs of Dongara West Australia for Singapore Headquartered Strategic Marine Pte Ltd with Nauti-Craft focussing on the hydraulic system, structural elements and electronic control of the suspension system. Electronic control is not covered in this paper.

Two previous papers presented at RINA conferences [1] and [2] provide detailed system descriptions so only a summary will be presented in this paper.

The suspension system provides heave stiffness, roll stiffness, pitch stiffness and near zero warp stiffness. The suspension also features "pitch coupling" where inputs at the front influence the response at the rear. All modes can be damped almost independently. The hulls are designed to follow the sea with the suspension absorbing impacts and controlling the chassis position and attitude. The system functions passively with controlled damping. Powered "roll-in" provides a mono-hull feel when cornering by tilting the deck into the turn, generating no lateral acceleration on the passengers. A powered deck attitude control system (DACS) allows the deck of the vessel to be controlled for near zero pitch and roll angles and alternatively allows the bow to be controlled to a fixed height. These features allow improved transit comfort (through reduced vertical accelerations) while increasing the operating profile (wave size and speed limitation) and the ability to more safely transfer to wind turbine pylons in larger seas. The ability of the vessel to operate in adverse conditions will reduce the cost of servicing off-shore wind turbines through increased vessel access availability.

The design and modelling processes are constantly improving and will continue to do so with more experience as this new technology is developed. The process can be summarised as:

- Production and updating of system models
- Validation of models
- Defining suspension arm hard points and hydraulic system layouts and components
- Generating limit and life load cases
- Component structural design

Typically, the system is modelled statically in Excel and dynamically in MSC.ADAMS [11] to produce hydraulic component sizes and system loads. Strength analysis is performed in Excel [12], NX NASTRAN [13] and PYTHON [14].

Where applicable DNV GL guidelines are followed but as this is a non-traditional technology much of the design is bespoke.

2. MODEL DEVELOPMENT (ADAMS)

2.1 HYDRAULIC SYSTEMS MODEL

The prime functions of the hydraulic models are substantially unchanged since the production of the previous paper [1]. However the models now simulate items relating to varying the operating mode of the vessel. Typically, these items are electrically operated isolation valves which provide a parasitic level of damping to the system when open and eliminate flow when closed.

Three independent models are used

- Pitch circuit model
- Roll circuit model
- Modal damping model (optional)

Parameters in these models represent the fluid/gas characteristics of real hydraulic components. True sizes are used throughout.

2.2 STRUCTURAL MODEL

This model generates the mechanical structure of the suspension system. Included are all the arms or links, the mechanical cylinder assemblies and all bearings. Bodies are created as rectangular prisms for the hulls and superstructure, the mass is assumed to be uniformly distributed for the rotational inertia properties but the centre of gravity is explicitly placed to match supplied data from the naval architect. Rotational inertias are reasonably correlated with Maxsurf Seakeeper [15] data for the complete vessel.

CAD models of the arms or links can be imported with true mass and inertia properties. Alternatively, simple geometry can be assigned the correct mass and inertia properties.

Hull and superstructure geometry is imported into the base craft model. While improving the model appearance, and ease of results interpretation, this geometry does not influence the analysis.

Bearings can be modelled as simple joints, e.g. spherical or hooke, etc., where appropriate but also can be implemented as bushes. Bushings are set up to provide 3-axis linear compliance and damping but are unconstrained for all rotations. Bushings allow load sharing in over constrained suspension systems.

All bushing and joint loads (forces) are available as measures. Each joint has 3-axis forces (x, y and z) expressed relative to both elements of the joint. This allows for simple force importation into finite element analysis (FEA) packages, calculation of basic torsion and bending modes and axial and radial bearing loads. Relative bearing angles are also available as measures.

Port and starboard are considered symmetrical.

2.3 COMPRESSION AND REBOUND STOP MODEL

Compression and rebound stops are modelled as springs utilising a spline fit of position vs generated force. Stops include a progressive soft stop and a linear, high stiffness, hard stop. The stops limit the suspension travel to the required stroke. Each stop can be positioned to act between any two parts.

2.4 HULL WATER INTERFACE MODEL

This model has been modified substantially since the previous paper [1].

Previously ADAMS contacts were used to generate the hull reaction forces on the water. This was a restricted approach due to the limited variables in the contact model. Secondly, the calculation effectively assumed the hull to have the same sectional width from the keel to the hull deck. Thirdly once the water part exceeded the hull deck height the section of hull in question would lose buoyancy. Fourthly added mass effects were not included.

The current model is set up as follows:

- The key geometry of the hull shape is derived from a CAD model. The keel line and chine height along with the hull beam are plotted against the distance from the bow.
- The hull is evenly divided between the bow and transom into equal length segments. Typically the number of segments is 25.
- The keel line, chine height and beam are estimated (linear fit) at the midpoint of each hull segment.
- For each hull segment a pre-simulation calculation is performed to generate the buoyancy force of the hull segment vs water height relative to the hull. This is converted into a spline.

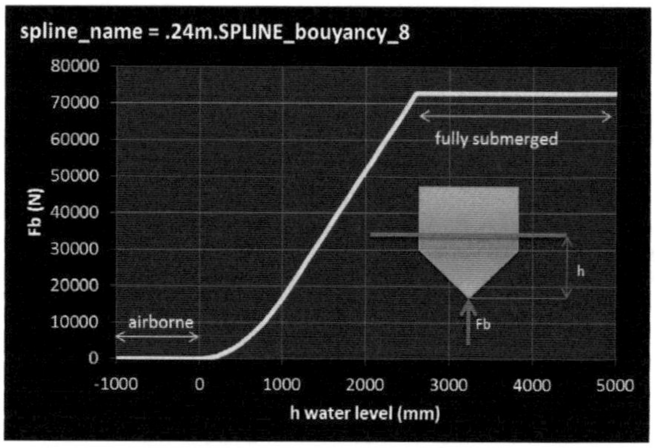

Figure 1: Hull segment buoyancy example

- The water is similarly split into segments along the length of the hull. Each water segment interacts permanently with the hull segment directly above it. A dummy segment is added fore and aft of the required segments for clarity when animations are produced. The height of each water segment is independently controlled by the method described below.
- The port and starboard hulls have independent sets of water acting on them.

The true vertical force between each segment of water and the hull part is calculated during the simulation in the following way (for each water segment):

- The water height, relative to segment's hull keel position is calculated. This is combined with the buoyancy spline to calculate the buoyancy force.
- The relative velocity is calculated. This is multiplied by a damping factor (currently linear) and multiplied by the buoyancy force. In this way the damping is proportional to the amount of hull in the water.
- The hull acceleration is calculated. This is multiplied by a mass effects factor (currently linear) and multiplied by the buoyancy force. A force calculated this way simulates the added mass of water moving with the hull. This mass varies based on the amount of hull in the water.
- The above forces are added for the total combined force.

These forces are applied to the appropriate hull at the segment midpoint locations and provide the inputs to the vessel.

An adjustment to the buoyancy force can be made that is dependent on vessel forward speed to crudely simulate the effects of planning. The current 24m WFSV is not expected to lift or change trim significantly in its operating speed range so this adjustment has been deactivated.

2.5 WAVE MODEL

Previously [1] the motion of the water segments was limited to a simple single frequency sine wave, for a continuous input, or stepped impulse functions. This is adequate for basic understanding but does not simulate real conditions.

The current model is based on the JONSWAP wave spectrum [3]. This model was considered most representative of the wave conditions likely to be encountered in the North Sea.

A relationship between wave period (T1) and significant wave height (Hs) was derived from Dogger bank data [4]. A linear fit was calculated from one month's worth of data.

The JONSWAP spectrum is calculated for a given Hs and calculated T1. The spectrum is calculated at a number of regularly spaced frequencies on a logarithmic scale in the range of interest from 0.5/T1 to 4/T1.

The component of wave speed along the vessels length is calculated at each frequency (based on the period) and this is combined with the forward velocity of the vessel to produce the encounter speed and hence encounter period or frequency. To simulate the progression of the wave along the hull the distance from the bow is divided by the encounter velocity to provide a time delay.

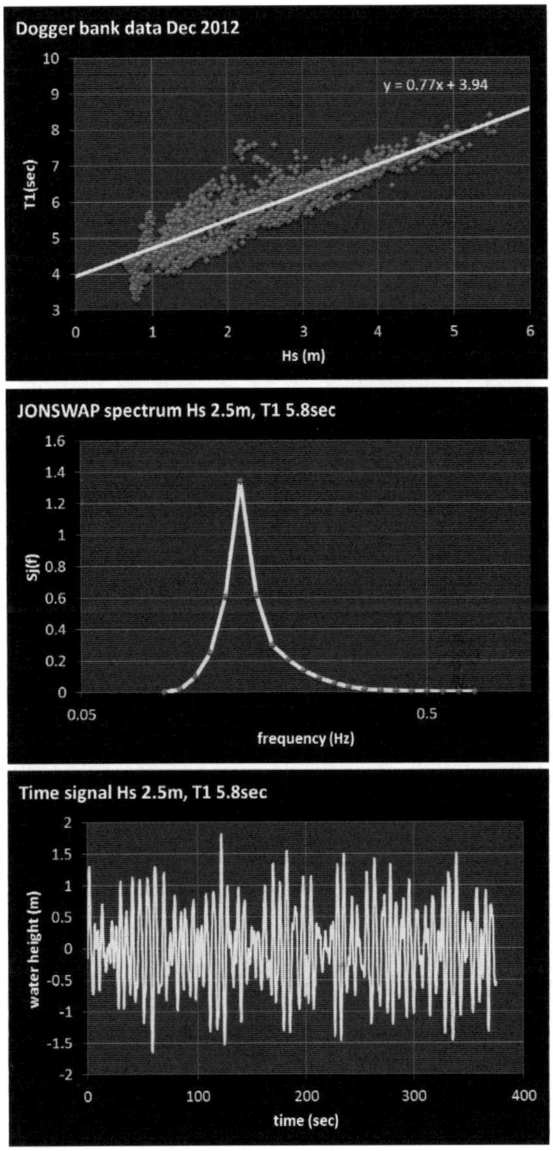

Figure 2: T1derivation from Hs, JONSWAP spectrum and resulting water height simulated time signal (mid ships for reference)

At each midpoint position along a hull a time series is calculated based on the addition of sine waves that provides the required spectrum. The wave position (height) is of the form:

$$x(t) = A_1.\sin(2.\pi.f_1.t + \Phi_1) + \ldots + A_n.\sin(2.\pi.f_n.t + \Phi_n) + \ldots + A_{20}.\sin(2.\pi.f_{20}.t + \Phi_{20}).$$

Where:

- f is the encounter frequency of the wave component
- t is the simulation time – the component time delay mentioned above

- The phase of each component Φ can be totally random and a different wave form is generated each time the model is run. However, for repeatability and comparison, a fixed table of phases is used. The wave shape generated repeats its structure every 628.2 seconds.

The Hs of this series has been measured (4 x standard deviation) and this correlates with the required Hs.

2.6 HULL WATER INTERFACE MODEL CALIBRATION/TESTING

2.6 (a) Buoyancy

The model was allowed to find equilibrium without any wave inputs. Comparison with Maxsurf [15] trim data supplied by the naval architect gave 0mm error at the transom and 120mm (high) 2000mm aft of the bow. This is equivalent to 55mm at mid ships (high) and 0.3 degrees trim (bow up).

2.6 (b) Added Mass

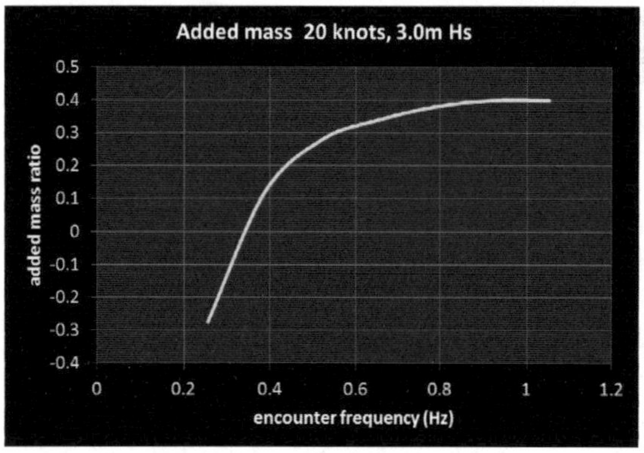

Figure 3: Maxsurf Seakeeper [15] data for heave added mass

At 20 knots head on and 3m Hs the prime encounter frequency is 0.27 Hz. The above data would indicate the added mass to be slightly negative. Considering the distribution of the encounter frequencies produced in the simulated wave model, and weighting higher frequency components as they are likely to be more strongly influenced by added mass, leads to an average encounter frequency 0.52 Hz. This provides an average added mass ratio of ~0.3.

2.6 (c) Water Damping

Reference vertical acceleration at the vessel's longitudinal centre of gravity was calculated using DNVGL HSLC Pt.3 Ch.1 Sec.2 2.2.2 [5]. Data used in this calculation mirrored the ADAMS model of the vessel. A sea of 3.38m Hs and 20 knots forward speed was used for the comparison. The DNV GL calculation makes no mention of the wave period so it is assumed to be in line with our linear estimate. The vertical acceleration was 1.72g.

The ADAMS model with the suspension locked was run in the same sea state at various water damping levels and the acceleration at the vessel's longitudinal centre of gravity was measured.

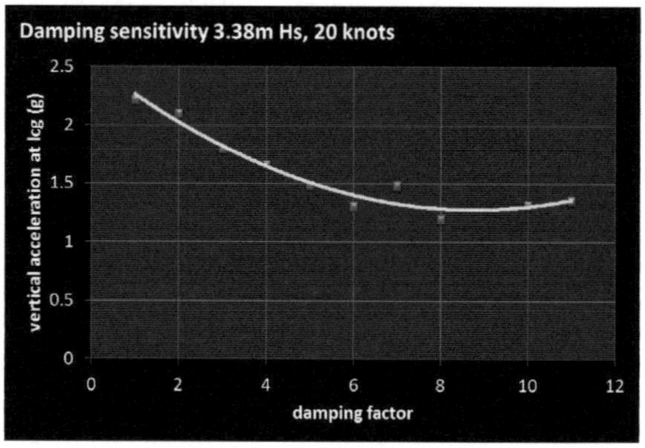

Figure 4: Acceleration vs water damping factor
A damping factor of 0.35 provided the required acceleration of 1.72g.

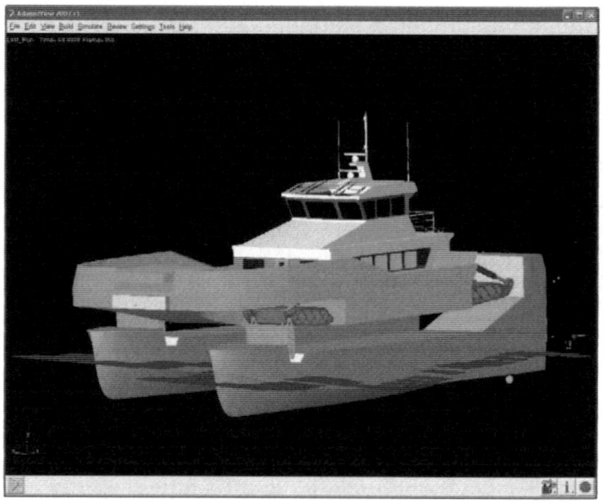

Figure 5: Full model: Bow quartering 26 knots 2.5m Hs

3. HYDRAULIC AND MECHANICAL HARD POINT DESIGN

The client's expectation of the WFSV is to provide much improved transit comfort and transfer performance. Active deck attitude control is also a specified requirement. These 3 criteria dictate the hydraulic system required. The design of the hydraulic system and the geometry of the arms/cylinders is iterative due to the complexity of the system and interactions.

Initially, expected masses and dimensions are entered into an Excel spreadsheet along with generic factors. These factors dictate expected strokes, mechanical advantages, maximum pressures, damping rates, buckling strengths etc. The model sizes hydraulic cylinders, accumulators, lines and dampers. Natural frequencies, pitch and roll rates are checked and the model reiterated if unreasonable.

The Excel model includes a simple geometric analysis of suspension arms and cylinders allowing the optimisation of hard points to provide linear mechanical advantages whilst ensuring cylinder lengths/strokes are physically possible.

The ADAMS model is assembled with the suspension hard points, hydraulic component sizes and static pressures derived from the Excel model. The hull shape is included based on data from the naval architect. This model is initially run to test stability of the system and confirm natural frequencies.

Personnel transfer is simulated and parameters such as bump stop characteristics, stroke, accumulator stiffness, hydraulic damping and hydraulic options, such as, secondary compliance are modified to achieve the required performance. Suspension arrangements can also be optimised. The use of a dropper between the hull and the front arm performs better than a dropper between the hull and the rear arm due to beneficial hull shunt. For the same engine thrust, typically the suspension equipped vessel does not slip at twice the Hs where the standard vessel slips.

Transit is simulated based on the operating profile described below in section 3.1. Various sea conditions, speeds and directions are simulated to test damper responses, bump stop characteristics, maximum pressures, etc.

Deck attitude controls modelled in Excel are tested to confirm power requirements, flows and required pressures.

Performance data of hydraulic components are sourced and entered into both Excel and ADAMS models. Tests are re-run and the hydraulic system/hard points are then defined.

4. LOADS PRODUCTION

Various load cases are produced via the ADAMS model. This data is processed before being used in bearing selection and finite element modelling of suspension components.

4.1 OPERATING PROFILE

Reverse engineering the approved operating profile of a current (traditional fixed hull catamaran) 24m WFSV by applying DNVGL HSLC Pt.3 Ch.1 Sec.2 2.2.2 [5] revealed the vertical peak acceleration at the longitudinal centre of gravity (lcg) was limited to 1g. At low speeds the Hs limit is a constant. Beyond this the 1g limit restricts the Hs.

The vessel was simulated across the estimated operating profile with the suspension locked (lockdown) and operating normally (ride control active). The damping in the suspension was optimised to allow maximum suspension travel without bump stop contact.

The longitudinal distribution of the peak vertical acceleration of the superstructure was determined. This was identified using two longitudinally spaced accelerometers placed on the superstructure part in the ADAMS model.

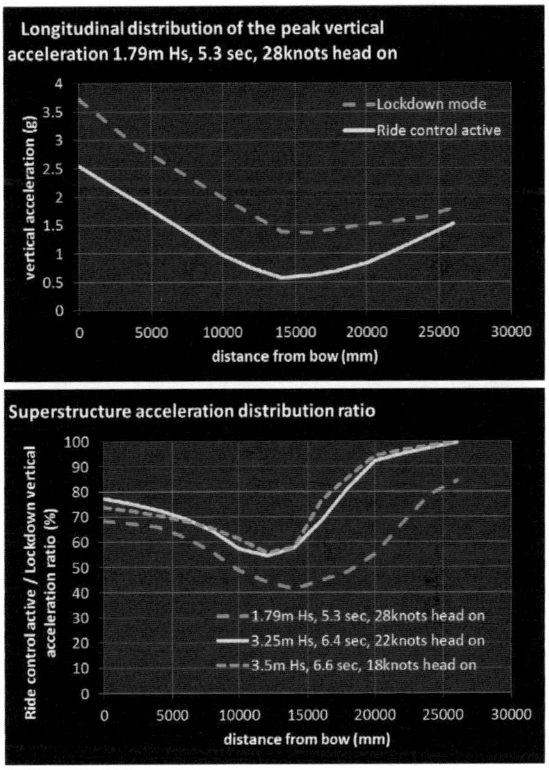

Figure 6: Typical acceleration distribution example of raw data and resulting acceleration comparison ratios

Ratios of ride control active acceleration levels to lockdown acceleration levels were calculated. Measured at the lcg of the vessel the accelerations are reduced to a maximum of 58% of lockdown levels. At speed the wave height is lower and the suspension damping can be reduced, the suspension performs better, and the ratio is 41%. It is also interesting to note that the accelerations are reduced significantly towards mid ships where passengers and crew are most likely to be positioned. The reduction in impact harshness and overall acceleration decreases passenger and crew fatigue and the motion sickness index.

Using the conservative value of 58% allows the production of a 1g limited operating profile. A significant wave height of 3m was selected by the client as a low speed limit. This attenuation may be increased with experience.

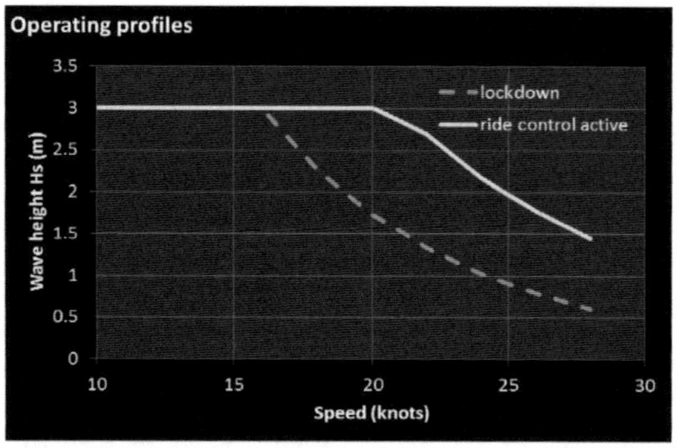

Figure 7: Operating profile

This (Figure 7) highlights the performance difference between a vessel fitted with the suspension system and effectively a standard vessel. This typically increases the vessel availability by 33%.

4.2 DUTY, LIFE CYCLE AND LIFE CASES

wave height range (Hs m)	Mid wave height (Hs m) for analysis	% of life cycle
Stationary		
0 to 0.5	0.25	2.6
0.5 to 1.0	0.75	13.7
1.0 to 1.5	1.25	17.3
1.5 to 2.0	1.75	14.9
2.0 to 2.5	2.25	9.1
2.5 to 3.0	2.75	5.4
Transit slow <10 knots		
0 to 0.5	0.25	1.0
0.5 to 1.0	0.75	5.0
1.0 to 1.5	1.25	6.3
1.5 to 2.0	1.75	5.5
2.0 to 2.5	2.25	3.3
2.5 to 3.0	2.75	2.0
Transit rapid		
0 to 0.5	0.25	0.6
0.5 to 1.0	0.75	3.0
1.0 to 1.5	1.25	3.8
1.5 to 2.0	1.75	3.3
2.0 to 2.5	2.25	2.0
2.5 to 3.0	2.75	1.2

An analysis of engine data of comparable WFSVs operating in the North Sea provided an estimate of the vessel usage profile.

ADAMS modelling of typical operating modes indicated loads generated while transferring, stationary and transiting in following seas were similar. Assuming the vessel transit is shared evenly between head seas and following seas leads to a reduced usage profile.

Dogger bank wave data [4] was analysed to determine wave height vs time distribution.

Combining the wave height distribution with the reduced usage profile leads to 18 life test cases each with an individual life in %.

Factoring in the operating profile and the ultimate vessel maximum speed, then these 18 test cases represent the complete load life of the vessel. Before finalising each test case, the suspension damping is optimised to minimise vertical acceleration. Transit rapid cases are in bow quartering seas. Typically, 1000sec of dynamic data is in each test case.

4.3 LIMIT LOADING CASES

4.3 (a) Longitudinal

DNVGL HSLC Pt.3 Ch.1 Sec.2 2.3.1 [5] provides a calculation for longitudinal g based on the peak vertical g at the vessel longitudinal centre of gravity. For typical worst case wave and speed conditions, based on the operating profile, a value of 0.29g is calculated. The longitudinal collision calculation DNVGL HSC Sec.2 2.3.1 [10] for the same vessel is 4.5g. The model is set at ride height. A load equivalent to 4.5g acting on the superstructure mass is reacted evenly by both hulls.

4.3 (b) Lateral

DNVGL HSLC Pt.3 Ch.1 Sec.2 2.3.2 [5] provides a calculation to estimate lateral acceleration. With additional terms added, the component of gravity aligned with the rolled deck is included and a value of 0.53g is calculated.

When cornering, "roll-in" control will roll the superstructure towards the centre of the turn and perform a coordinated turn. In this state, the loads on the inner and outer hulls are identical and little speed is lost. Passengers should be relatively unaware of the turn. Typically, the suspension travel will allow a 0.4g turn compared to 0.2g for a standard vessel. In this case the suspension on the outside of the turn is extended providing an increased moment arm leading to added forces on the suspension arms. A reasonable limit case is 0.5g with the suspension extended to bump stop contact.

Figure 8: "Roll-in" static test

Both the longitudinal and lateral limit cases are modelled statically (at equilibrium) providing one load case each.

4.3 (c) High speed transit

Through multiple tests it has been proven that the peak suspension loads are generated in the area where the operational profile transfers from being wave height limited to acceleration limited. Two dynamic limit test cases are run at the Hs limit in head seas and bow quartering seas and two test cases are run at the wave limit of the life test case, again in head seas and bow quartering seas.

4.4 POST PROCESSING

All the dynamic data is automatically tested for unrealistic spikes caused by ADAMS solver inaccuracies. This is only an issue if the solver accuracy is slackened.

Secondly the data produced in the dynamic limit tests effectively repeats every 628 seconds. While this represents > 150 wave encounters there is a chance that in the life time of the vessel the loads calculated could be exceeded. DNV GL has specified a probability of exceedance of 1e-8. Based on prior knowledge, torsion around the longitudinal axis, on the front swing arm is critical to the design. This moment was calculated at each time step from the bearing loads and analysed for a best fit to a statistical distribution. This distribution then defines the design moment required with the correct probability of exceedance. The net result is a single factor applied to all of the original loads to generate the factored design loads.

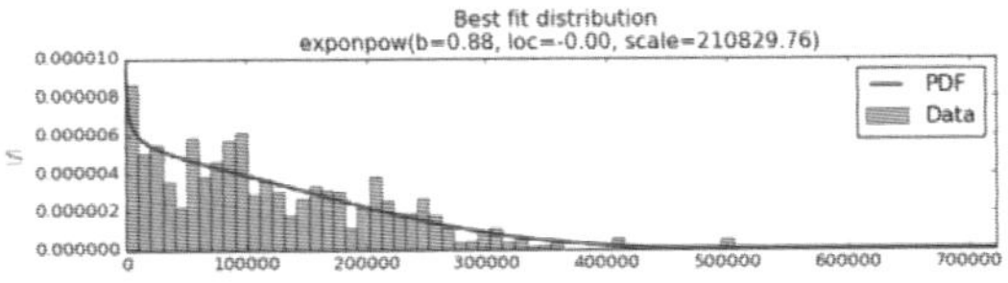

Figure 9: Distribution fit for probability of exceedance calculation

4.5 DATA REDUCTION

The post processed dynamic limit time data is searched to find the peak loads in the following cases for each of the major suspension components (front arm, front dropper and rear arm):

- Component
  - torsion
  - bending, 2 axes
  - shear, 2 axes
- bearing location forces
  - magnitude
  - x, y and z
  - x, y x, z y, z
  - radial
  - axial

This could potentially extract 59 limit load cases for each component but many of the cases occur at the same time step. Typically, ~18 unique limit load cases will be generated per component.

These limit load cases together with the two static limit cases are combined and referred to as limit load cases.
The spike tested dynamic limit data is also searched along with the two static limit cases to find the peak loads in multiple cases for each of the bearing assemblies. This returns data for bending and shear etc. in the clevis webs, bearing tangs and pins. The orientation of each of the webs and tangs is combined into the calculation to ensure the data is rotated to the correct axis system. This data is referred to as the clevis limit load cases.

Each of the 18 spike tested life test cases is reduced to find the following data for each bearing (Dynamic bearing data):

- Dominant frequency (via FFT)
- Total oscillation angle (maximum angle – minimum angle)
- Maximum radial force
- Maximum axial force

Further each life load test case is searched for two load cases representing maximum and minimum arm torsion.

5. COMPONENT STRUCTURAL DESIGN

The structural design aims to minimise weight and maintain appropriate minimum safety factors.

5.1 CLEVIS, PIN AND TANG SIZING

The code ASME BTH [6] is used to calculate all the clevis, pin and tang dimensions. The code is conservative so the clevis limit load cases can be used directly without any factoring. The safety factors include an allowance for fatigue.

5.2 BEARING SIZING

The bearings are sized initially based on the pin size. Each of the 18 dynamic bearing data sets is entered into an excel calculation based on SKF plain spherical bearing life [7]. The % life and the resulting calculated life are combined using a form of the Miner's rule [8] to produce the total bearing life in hours. The minimum design life is 5 years at 3000 hours per year.

The bearings are also tested for all the limit load cases (radial and axial loads) for a static safety factor > 1.

Should the bearing not meet life or static limits then the pin size is upgraded and the bearing recalculated.

5.3 ARM DESIGN

The arms (front arm, front dropper and rear arm) are modelled in Solid Works and transferred into NX NASTRAN for finite element analysis (FEA).

5.3 (a) Limit design

Material allowable stress [9] is used with a safety factor of 1. To be confident of not exceeding the allowable stress the factored limit load cases are tested.

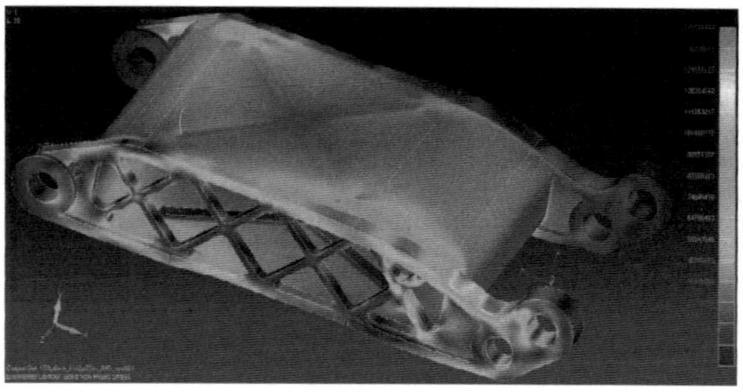

Figure 10: Stress result from FEA limit testing

5.3 (b) Fatigue design

As previously mentioned, arm torsion is driving arm stress. This analysis assumes arm torsion is proportional to arm stress.

The analysis is as follows for each arm:

- for each life test case (18x)
    - perform FEA at T max
    - perform FEA at T min
    - transfer resulting stress distributions to PYTHON
    - for each element (600,000x)
        - find stress max and stress min
        - scale against T max and T min and produce a time series of stress
        - rain flow count stress swing
        - combine with S/N curve
        - calculate life with Miner's rule and store data
    - next element
- next life test case
- for each element
    - recall stored life data
    - calculate overall life with Miner's rule, combining life test case life % and element life in each life test case
- next element
- find element with minimum life and return as fatigue life of the structure

The arm design is modified until it passes the required fatigue life.

6. CONCLUSIONS

The engineering design process has been evolving as more knowledge is gained through modelling, design and testing of the Nauti-Craft suspension system. The steps and methodology described here provide a robust system of design within the limits of current knowledge. The process has been provisionally approved by DNV GL. Once the 24m WFSV is constructed and tested the design process and modelling will be updated providing further optimization of weight and cost.

7. ACKNOWLEDGEMENTS

Nauti-Craft acknowledges:

- The Carbon Trust (UK) for its support through the Offshore Wind Accelerator (OWA) program.
- Strategic Marine Pte Ltd
- Southerly Designs

8. REFERENCES

1. LONGMAN, M., Nauti-Craft Pty Ltd, 'DEVELOPMENTS IN COMPUTER MODELLING OF ATTITUDE AND RIDE CONTROL SYSTEMS FOR A SUSPENDED MULTI-HULLED WINDFARM VESSEL', *The Royal Institution of Naval Architects, Design & Operation of Wind Farm Support Vessels, 29-30 January 2014*

2. ROBERTSON, A.J., MONK, R., Nauti-Craft Pty Ltd, 'A REVOLUTIONARY RIDE CONTROL SYSTEM FOR MULTI-HULL HIGH SPEED MARINE VESSELS, *RINA High Speed Marine Vessels 2-3 March 2011 Fremantle*

3. DNV Article 2.8 - Wave Spectra: Section 2.8.3 - JONSWAP Spectrum.

4. East Site Dogger Bank Wave Monitoring (Dec-12)

5. DNVGL HSLC Pt.3 Ch.1

6. Design of Below-the-Hook Lifting Devices ASME BTH, ASME 2014 ISBN: 9780791869444

7. http://www.skf.com/binary/96-122020/6116_1-EN.pdf

8. Palmgren-Miner linear damage hypothesis, https://en.wikipedia.org/wiki/Fatigue_(material)#Miner.27s_Rule

9. DNVGL-RU-HSLC-Pt 3 Ch 3, Allowable stress Aluminium

10. DNVGL HSC Sec.2 2.3.1, accommodation and escape measures: design acceleration levels

11. ADAMS, MSC Software Corporation

12. MS Excel, Microsoft Corporation

13. NX NASTRAN, Siemens PLM Software

14. PYTHON, Python software foundation

15. Maxsurf and Maxsurf Seakeeper, Bentley Systems

9. AUTHORS BIOGRAPHY

Mike Longman holds the current position of Chief Engineer at Nauti-Craft Pty Ltd. He is responsible for design, modelling and testing of suspension systems. His previous experience includes concept development, design, modelling, manufacturing, commissioning, testing and production development of automotive interconnected suspension systems with Kinetic suspension systems. This experience has been adapted to the production of prototype water based craft.

ACCESS TO A FLOATING WIND TURBINE

M Shanley, C S Wright, C Desmond, A Otter, J Murphy, Lir NOTF, MaREI Centre, ERI, University College Cork, Ireland

SUMMARY

The offshore wind turbine service industry is now well established with a large number of turbines being successfully operated and maintained. A number of methods and technologies are available to allow the safe transfer of service crews to these primarily fixed monopile installations. The most common of these is the bow transfer method which uses a combination of a high friction fender and a large vessel thrust to minimise relative motion between the bow and the turbine foundation.

An upcoming challenge for the offshore wind turbine service industry will be the increasing use of floating foundations in far offshore and deep water sites. A number of structures are currently being developed and the first commercial floating wind farm is expected to be commissioned in late 2017. The use of floating structures will make it more difficult to ensure crew safety and comfort during transfer operations as the interaction between two floating bodies needs to be considered. Thus, the bow transfer method used to access fixed foundations may not be suitable for accessing floating turbine platforms.

This paper will use a combination of physical and numerical modelling to assess the ability of a wind farm service vessel to maintain contact with a floating offshore wind turbine structure by use of the bow transfer method.

1. INTRODUCTION

Operations and Maintenance (O&M) of an offshore wind farm is substantially more expensive than for an onshore wind farm, it is vital that accessibility limits are known so as to aid planning of the operation.

An assessment of the current state of the art of wind farm service vessels (WFSVs) can be made by reviewing the database of 498 vessels provided by 4C Offshore Ltd, a leading consultancy and market research organisation targeting the offshore energy industry [1]. The vast majority of the vessels listed in that database are high speed catamarans with a cruising speed of 15-25 knots and are generally between 15-24m in length. These vessels can carry a cargo in the range of 3-15t and are typically of aluminium construction, though glass reinforced plastic and other composites are used.

The crew transfer method of vessels is not listed in the 4C Offshore database. Catamarans of the type listed generally use the industry standard bow transfer method described in the summary. A large bollard pull allows the connection between the WFSV and the turbine to be maintained safely for most wave spectra with a significant wave height (Hs) of up to 1.5m [2,3,4,5]. It has been suggested that for the step across transfer the relative motion between the point on the vessel and the turbine docking poles should be essentially zero. [4]. However, a recent industry review showed that crew transfer vessels can safely transfer teams in sea states with Hs of up to 1.8m Hs. [6]

Access to fixed wind turbines can be achieved at higher sea states by use of alternative vessels designs such as the Natilia Bekker small-waterplane-area-twin-hull (SWATH) vessel which can access wind farms at a 2.5m Hs [7]. Another possibility is the use of walk to work systems such as the MaXccess or Ampelmann devices which can be installed on larger service vessels. An Ampelmann installed on a 70m long vessel has allowed safe access to fixed offshore wind turbine in a seastate with a Hs of 3m [8]. The Ampelmann system compensates for the relative motions between the vessel and the turbine, transfer of crew is permitted once relative motions between the stabilised platform and turbine is less than 0.5m in heave [9].

With the advent of floating wind turbines, the problem of access is compounded by the motion of the turbine platform. If using the bow transfer method, excessive displacements caused by loss of frictional contact between the WFSV and the turbine platform may create a serious incident if a transfer operation was being carried out.

Numerical and physical modelling of the interaction of a WFSV with an offshore wind turbine, both fixed and floating, has been used in a number of studies to better understand these interactions. Numerical modelling studies have examined the problem using time and frequency domain modelling and have used a variety of methods to account for the frictional contact between the WFSV and the wind turbine tower.

In one study [10], a numerical method of predicting whether or not a slip will occur was developed by using a qusai-static and dynamic model to describe the system and the Coulomb frictional relationship at the contact point. The major forces of the system were accounted for and a static and dynamic analysis were carried out in the time domain focusing on whether or not a slip would occur. It was found that slips generally occur in the positive Z direction and that the coefficient of friction is

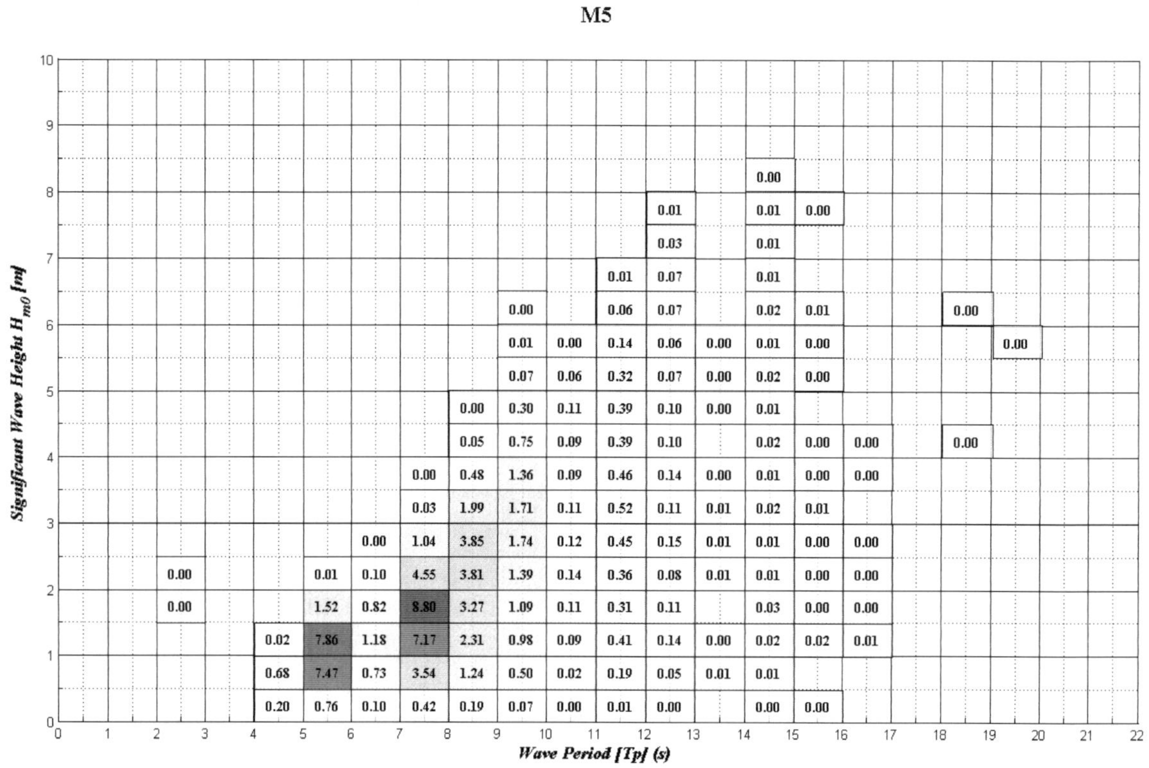

Figure 1: M5 Scatter Plot

of significant importance. This study considered a fixed turbine foundation.

In another numerical study [11], a statistical analysis of the docking procedure to a fixed foundation was examined. The bow transfer involved estimating the coefficient of friction and tolerance of the standard deviation of the slips. When combined with the sea state, estimations of accessibility were determined. It was found that upwards slips were more likely to occur than downwards slips. In addition, the peakedness of the spectrum was found to have a negligible effect on the performance assessment.

In terms of physical modelling in this area, a series of experiments were carried out using self-propelled generic catamaran models [3,4,5]. Scale model tests of a 24m WFSV with an active motion compensated gangway were carried out and the vertical forces recorded [3]. It was noted that performance was limited more by stern swamping and propulsion ventilation in stern seas and severe roll in beam seas rather than exceedance of the thresholds of the access system [3]. It was found that for head seas that slips were more common in the positive Z direction. In addition, it was noted that the efflux from the propellers caused an increase in wave steepness, which in turn caused a greater variation in longitudinal force, and consequently slips [5].

Both numerical and experimental investigation of a WFSV at a fixed offshore wind turbine were examined in [12]. In the physical testing, the vessel fender was attached to the WFSV using load cells measuring axial and tangential force. Fender friction was also considered in a separate series of dry tests. In the numerical model, a time domain simulation was created considering the Froude-Krylov forces and forces on the catamaran in addition to the diffraction due to the monopile. The friction contact was modelled as Coulomb friction. Good agreement was found between the numerical and physical simulation [12] and it was shown that the numerical model could accurately determine the risk of a slip occurring. In a follow up paper by the same author [13], the numerical model was further refined with a hyperelastic material used to describe the mechanical behaviour of the fender which was analysed using the finite element method (FEM).

The study discussed previously examined access to a fixed turbine foundation. In terms of access to a floating turbine, a numerical study was presented in [14,15,16]. A two body frequency domain BEM model was used with modelling of the contact point as fixed. The forces at the fixed contact point were recorded during simulations and analysed to determine the required frictional force to maintain contact. The mooring forces of the floating wind turbine were linearised as the entire model was solved using linear systems in the frequency domain. It was found that the wake of the floating wind turbine and the relative motion between the two bodies had significant effects on transfer. Some studies relating to fixed wind turbines are outlined in the following paragraphs.

In this paper, results from both a numerical and physical test campaign will be presented in which WFSV access to a floating wind turbine will be considered.

2. PHYSICAL MODELLING

The configuration used in the physical test campaign is described in this section.

2.1. TEST FACILITY

Testing was conducted in the Deep Ocean Basin of the Lir National Ocean Test Facility, Ringaskiddy, Cork, Ireland. The basin, has dimensions of 35 m long x 12 m wide x 4 m deep and is equipped with 16 hinged force feedback paddles capable of a peak wave generation condition of Hs = 0.6m, Tp = 2.7s and Hmax = 1.1m. A movable floor plate allows the water depth to be adjusted to a maximum of 3 m. During testing at a scale of 1:30, the tank water depth was set to 3 m in order to simulate a water depth of 90 m at full scale.

Wave height in the tank was monitored by use of 6 No. 1m conductive wave probes. The probes were calibrated prior to testing and found to have a measurement accuracy of +/- 2 mm.

In order to track the motions of models in the Deep Ocean Basin, a system provided by the Swedish company Qualisys is used. The installation consists of four Opus 3-series cameras, data from which are captured at a rate of 32 Hz and processed using the Qualisys Track Manager (QTM) software.

The Qualisys system requires the installation of reflective markers on all floating models in order to define rigid bodies, the motions of which are tracked in the six degrees of freedom.

2.2. SERVICE VESSEL

The catamaran was chosen to be representative of a larger type of catamaran vessel such as the Windcats 101, which have a maximum thrust of 19.5t [17]. Thus a catamaran 28.8m long with a draught of 1.2m and a thrust of 20t was studied. The trust was achieved through two lines from the stern of the model as can be seen in Figure 2c. The fender was constructed from rubber. The boat landing was constructed from 10mm (model scale) aluminium pipes protruding 50mm (at model scale) from the face of the stern outer column of the floating wind platform as can be seen in Figure 2a. The overall model setup is shown in Figure 2b.

Figure 2a: Physical Model Setup

Figure 2b: Physical Model Setup

Figure 2c: Physical Model Setup

2.3. FLOATING WIND TURBINE PLATFORM

The floating wind turbine platform considered was the Hexwind Tension Leg Platform (TLP) described in [18]. This platform has been designed to support the NREL 5 MW reference turbine [19]. Platform, tower and rotor nacelle assembly (RNA) mass combined was 221,5440 Kg, Columns are 9m (centre) and 6m (outer), displaced volume was 3911.5 m^3. No wind loading was applied during testing and the RNA mass was applied at the full scale RNA centre of gravity (CoG).

2.4. SEA STATES

The wave conditions considered in the analysis were selected to be representative of the M5 buoy location. A potential deep water site in the Celtic Sea. The measured scatter plot is shown in Figure 1 whilst the location of the buoy is shown in Figure 3.

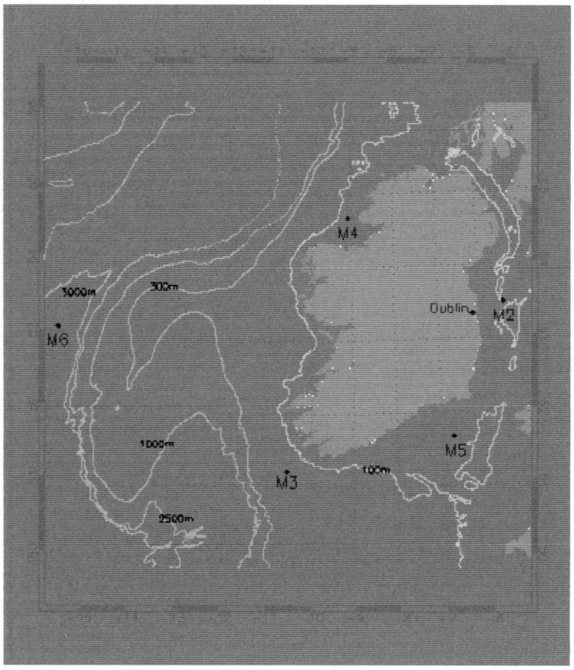

Figure 3: Irish Weather Buoy Network

For each sea-state, 124 random seed wave timeseries (TS) were generated at one-hour full scale duration. The time used in the experiment ~8.5mins model scale (discounting 1 minute start up time) was analysed on a zero up crossing wave by wave basis. The Hmax of each TS was calculated. The Anomaly Index (Ai = Hmax/Hs) of each TS was calculated. The TS with Ai closest to that specified, Ai = 1.6, 1.8, 2, was chosen for experimental testing in the basin. Waves were then Froude scaled to the basin.

Table 1: Irregular Waves at Full Scale

	Hs (m)	Tp (s)	fp (Hz)	Water Depth (m)
1	0.75	5.5	0.182	90
2	0.75	7.5	0.133	90
3	1	5.5	0.182	90
4	1	7.5	0.133	90
5	1.25	5.5	0.182	90
6	1.25	7.5	0.133	90
7	1.5	5.5	0.182	90
8	1.5	7.5	0.133	90
9	1.75	7.5	0.133	90
10	2	7.5	0.133	90

The actual Ai calculated for each sea state is given in Table 2.

Table 2: Anomaly Index (Ai = Hmax/Hs) for Numerical Model

No.	Ai = 1.6	1.8	2.0
1	1.599	1.795	2.006
2	1.599	1.791	1.981
3	1.599	1.795	2.006
4	1.599	1.791	1.981
5	1.599	1.795	2.006
6	1.599	1.791	1.981
7	1.599	1.795	2.006
8	1.599	1.791	1.981
9	1.599	1.791	1.981
10	1.599	1.791	1.981

3. NUMERICAL MODELLING

Hydrodynamic parameters were computed using the potential flow theory radiation/diffraction solver ANSYS AQWA for 12 wave directions (30° intervals) and 50 wave frequencies from 3.2s- 30s. The mesh defeaturing tolerance and maximum element size is set as 1.6m and 0.8m respectively for Hexwind and 1.0m and 0.3m for the catamaran.

Time domain analysis was conducted using OrcaFlex v10.1a. Viscous drag forces were applied to the platform using modified drag only Morison's equation elements, using drag coefficients from [20]. The mooring tendons were modelled as Morison's equation elements, discretized into twenty sections each approximately 5m long. Tendon axial stiffness was 10E6 kN.

As the flexible WT tower for tension moored towers influence the pitch natural frequency the flexible tower was numerically modelled using line elements. The catamaran was connected to Hexwind using a hinge connection which only allows relative pitch between the two bodies. Vessel thrust was modelled as a constant force applied at the same location as the experimental model. No wind and current loads were applied.

All simulations were conducted on a single 64-bit desktop PC with 2 x 3.40 GHz Intel i7-4770 processors and 12.0 GB of RAM. The simulation processing time to simulation run time ratio was 1.17.

4. RESULTS & DISCUSSION

In this section, a comparison between the numerical and physical test results is presented in order to assess the performance of the bow transfer method for a floating wind turbine.

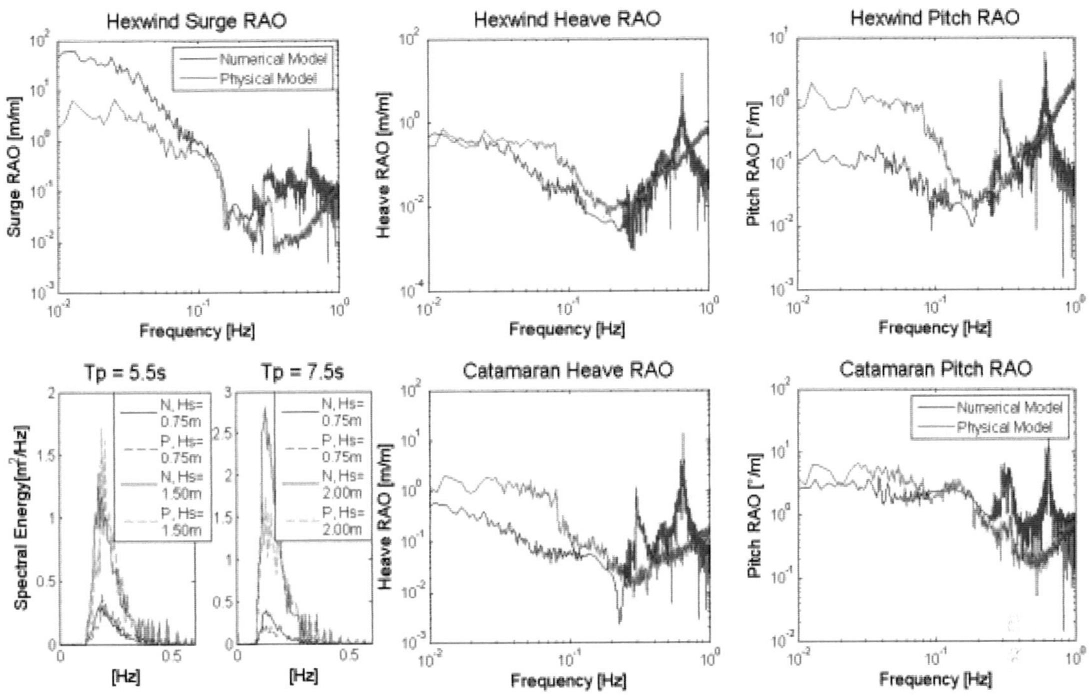

Figure 4: RAO and Wave Spectra Data

4.1. RAO COMPARISON

Firstly, the response amplitude operators (RAOs) determined for the two floating bodies, both numerically and physically, will be examined. An RAO is the ratio of a floating body's motion to the input wave height and is a critical parameter used to assess the sea keeping behaviour. This study uses the cross-spectral auto-spectral method [21] to calculate the RAO.

$$RAO = H(\omega) = \frac{S_{xy}(\omega)}{S_{xx}(\omega)}$$

where, $H(\omega)$ is the frequency response function, $Sxy(\omega)$ and $Sxx(\omega)$ are the cross-spectral and auto-spectral densities of the input $x(t)$ and the output $y(t)$, in the frequency domain, respectively. A comparison of the RAOs obtain for the two floating bodies by both numerical and physical methods is shown in Figure 3 along with an assessment of the actual wave spectra produced in the wave basin during testing for two selected cases.

As can be seen in Figure 4, the energy content of the waves produced during testing is lower than expected for the Tp = 7.5s and slightly too high for the Tp = 5.5s.

In terms of RAOs, for the Hexwind platform in Surge we see that the numerical and physical model agree well for the wave excitation frequencies especially in the range of 10 – 5s. The numerical model is seen to over predict response close to the natural frequency of the platform which would indicate that the drag coefficients used are too small. This discrepancy could also be explained by the fact that the motions of the physical model are damped by the six tendon load cell cables.

In heave for the Hexwind platform, we see that the numerical model generally under predicts, apart from the exaggerated high frequency peak. Possible reasons for this peak include the fact that the numerical model has undamped hydrodynamic coupling, the discrepancy between the flexible (numerical) and rigid (physical) tower or perhaps oscillation of the mooring tendons.

For pitch in the Hexwind platform, the numerical model generally under predicts, apart from two high frequency peaks. Possible explanations for this peak include the fact that the numerical model does not include sum frequency loads which are dominant in pitch along with the possible reasons outlined for discrepancies in heave.

In terms of the RAO for the catamaran, in heave the rigid hinge coupling in the numerical model results in heave motion being a product of the pitch & heave of the Hexwind model and of the and catamaran pitch. The numerical modelling under predicts response, which indicates that the hydrodynamic interaction could be negatively affecting results, or that reflections dominate results. The two high frequency components in the numerical model appear to be linked to the Hexwind pitch motion.

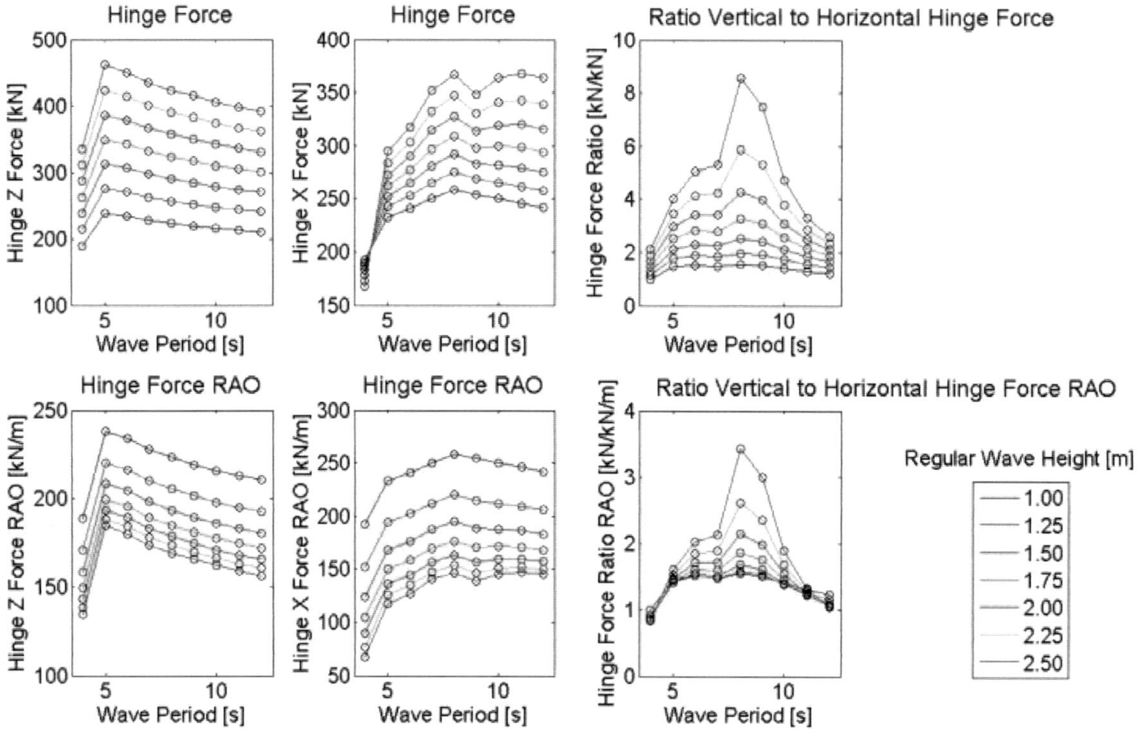

Figure 5: Numerically Modelled Hinge Forces

In pitch the RAO for the catamaran are seen to agrees well Numerical and physical model match well especially well from 12 – 4s. Two high frequency components exist in the numerical model which appear seem to be related to the Hexwind pitch motions. In general terms, the agreement between the numerical and physical RAO is considered sufficient to allow analysis.

4.2. WFSV / TURBINE INTERACTION

The forces between the WFSV and the wind turbine platform were not measured during physical testing. In Figure 5, the numerically modelled hinge forces are presented. The hinge force is the contact point between the bow fender and the boat landing as described in Section 3.

The peak in the Z force at 5 seconds relates to the catamarans natural pitch period occuring at this period. A significant portion of this force is absorbed by the flexible, compressible rubber fender and smaller slips that do not show up in the analysis. The numerical model is seen to capture the non-linear behaviour of the interacting bodies, as the RAO's of each hinge force demonstrate.

Experiments with irregular wave numbers 8, 9 and 10 with an Ai = 2.0 could not be completed due to the catamaran yawing off the boat landing. This phenomenon was believed to be due to experimental set, the large vessel motion and number of slips occurring. Attempts to constrain the vessel in sway and thus yaw, failed as slippage was also constrained.

In Figure 6, a comparison between the risk of slippage between the bow fender and the floating platform is provided. Results are presented as a probability of exceedance of relative motions in the vertical direction Z.

Results are presented for three Ai's for each the 10 sea states summarised in Table 1.

It is indicated in Figure 6 that the numerical model over predicts motions at small wave heights and also then under predict when slips occur. Generally, it is seen that the hinge force seems to be quite significantly over predicted numerically. During the physical tank testing with a bollard pull of 200 kN, and coefficient of friction of approximately 0.8, no slippage was observed for 1.5m regular waves, whilst an increase to 1.75m cause significant slipping. The numerical model predicts that for wave heights of 1.5m force a force of 200 - 300 kN would cause significant slippage.

In Figure 7, the confidence rate of slips not occurring as a percentage of the number of zero crossings is plotted. The confidence rate of slips increases as the vertical movement defining a slip is increased. It can be seen that by defining a slip event as a movement of the fender by 0.1m in the z direction then as the anomaly index

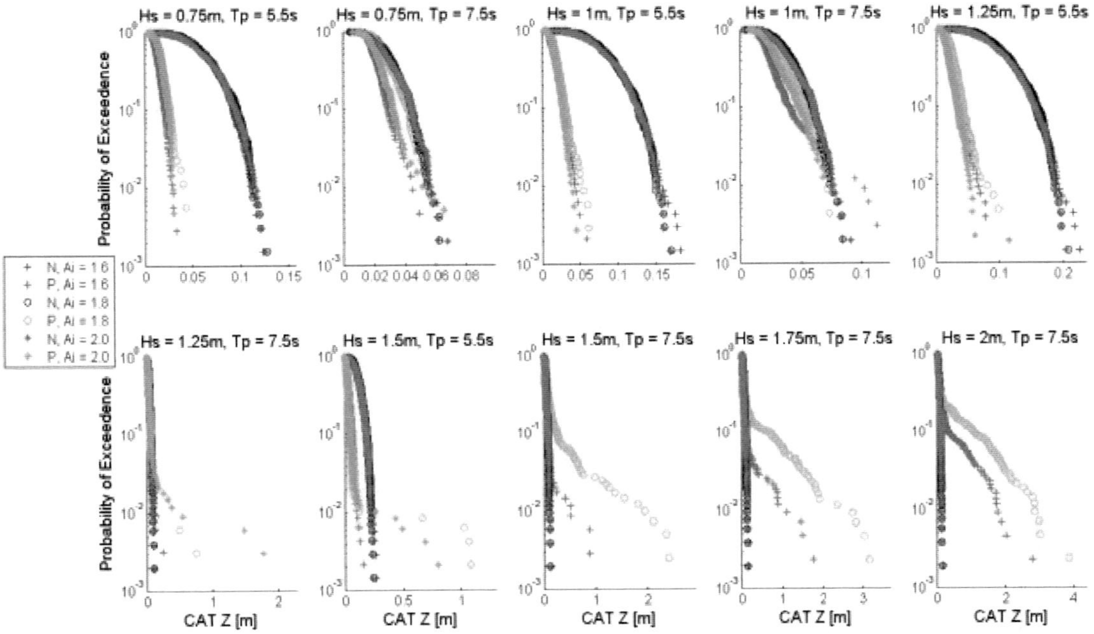

Figure 6: Probability of Exceedance of Relative Vertical Motions

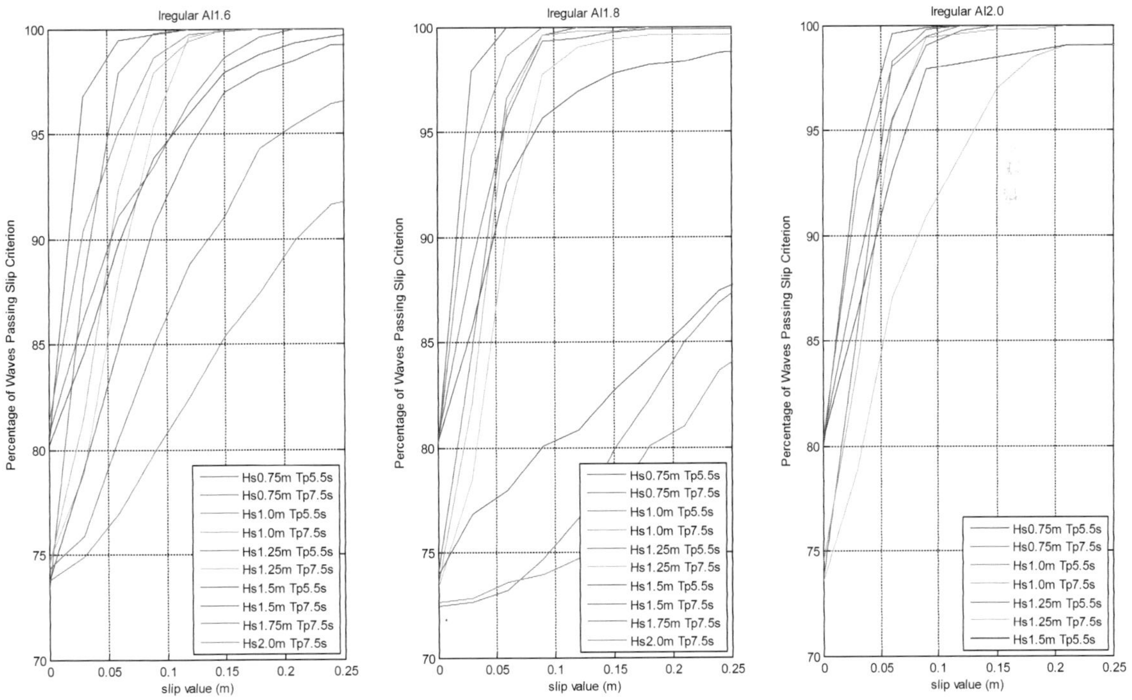

Figure 7: Confidence Rate of Slips Not Occurring as a Percentage of the number of zero crossings

increases from 1.6 to 2.0 a confidence rate of 95% suggests the possibility of safe transfers occurring for a 1.5m Hs reducing as the anomaly index increases. Sea states with a 1.75m and 2.0m Hs show a low confidence rate of safe transfers for all cases.

5. CONCLUSIONS

The numerical and physical modelling of a catamaran WFSV docking with a TLP wind turbine were presented in this paper. The response of the interaction between these bodies was studied. The discrepancy between the numerical and physical model showed the requirement for using physical model test to validate the numerical model. The physical model showed the highly non-linear behaviour of the system, which justifies the use of a time-domain model incorporating some non-linearities. The anomaly index has a large effect on the safe transfer limits.

Future work will investigate; the numerical/physical discrepancies, the use of flexible tower during testing, the effect of varying thrust / friction of rubber in testing, measure bow force in testing, and study varying wave incident angles.

6. ACKNOWLEDGEMENTS

The Authors wish to acknowledge Lir National Ocean Test Facility and the MaREI research centre, for supporting the study.

7. REFERENCES

1. 4C Offshore. Wind farm service vessels. http://www.4coffshore.com/windfarms/vesselcategorywindfarmservicevcid2.html, 2015.[Date accessed: 17/12/2016].

2. S. Livaniou, S. Iordanis, P. Anaxagorou, B. Mocanu, R. Sykes, J. Goormachtigh, J.M. Christensen, J.T. Kristensen, and M. Antrobus. Key design parameters and criteria related to installation and maintenance vessels design; their layouts, crane operations and access systems D.3.2. http://www.leanwind.eu/wp-content/uploads/GA_614020_LEANWIND_D3.2.pdf, 2015. [Date accessed: 04/08/2015].

3. S. Phillips, I.B. Shin, C. Armstrong, and D. Kyle-Spearman. Performance evaluation of wind farm support vessels. In Design and Operation of Wind Farm Support Vessels, pages 141–145, London, United Kingdom, 2014. Seaspeed Marine Consulting Ltd, The Royal Institution of Naval Architects.

4. S. Phillips, I.B. Shin, and C. Armstrong. Crew transfer vessel performance evaluation. In Design and Operation of Wind Farm Support Vessels, pages 29–33, London, United Kingdom, 2015. Seaspeed Marine Consulting Ltd, The Royal Institution of Naval Architects.

5. H. Maclean, C. Armstrong, I. Shin, and S. Phillips. Fast crew transfer vessels - transit and transfer performance research. In Design and Construction of Wind Farm Support Vessels, pages 19–24, London, United Kingdom, 2016. Seaspeed Marine Consulting Ltd, The Royal Institution of Naval Architects.

6. S. Shenton. Approaching a critical review: an assessment of real world crew transfer vessel capabilities at Gwynt y Môr https://ore.catapult.org.uk/wp-content/uploads/2016/05/Approaching-a-critical-review.-An-assessment-of-real-world-crew-transfer-vessel-capabilities-at-Gwynt-y-M%C3%B4r.pdf 2016. [Date Accessed: 2017-02-27]

7. Bard. The first special vessel for personnel transfer in offshore wind farms. http://www.bard-offshore.de, 2014. [Date accessed: 20/10/2015].

8. D.J.C. Salzmann. Ampelmann Development of the Access System for Offshore Wind Turbines. PhD thesis, 2010.

9. We@Sea. Ampelmann; a motion-compensating platform for accessing wind turbines. http://www.we-at-sea.org/wp-content/uploads/2013/01/1512004-012_summary.pdf, 2013. [Date accessed: 19/10/2015].

10. T. Josse, A. Billet, and 2011 Leen, S.B. Prediction of supply vessel motion during transfer to a fixed structure. In ASME 2011 30th International Conference on Ocean, Offshore and Arctic Engineering. American Society of Mechanical Engineers, 2011.

11. M.K.Wu. Numerical analysis of docking operation between service vessels and offshore wind turbines. Ocean Engineering, 91:379–388, 2014.

12. D.F. Gonz´alez, M. Lemmerhirt, M. Abdel-Maksoud, M. K¨onig, and A. D¨uster. Numerical and experimental investigation regarding the landing manoeuvre of a catamaran vessel at an offshore wind turbine in waves. In ASME 2015 34th International Conference on Ocean, Offshore and Arctic Engineering. American Society of Mechanical Engineers, 2015.

13. M. K¨onig, A. D¨uster, D.F. Gonz´alez, and M. Abdel-Maksoud. Simulation of safety-relevant situations regarding the interaction of service

ships with offshore wind turbine plants. In ASME 2015 34th International Conference on Ocean, Offshore and Arctic Engineering, pages V007T06A044–V007T06A044. American Society of Mechanical Engineers, 2015.

14. Michele Martini, Alfonso Jurado, Ra´ul Guanche, and I.J. Losada. Evaluation of walk-to-work accessibility for a floating wind turbine. In ASME 2016 35th International Conference on Ocean, Offshore and Arctic Engineering, pages V006T09A038–V006T09A038. American Society of Mechanical Engineers, 2016.

15. R. Guanche, M. Martini, A. Jurado, and I.J. Losada. Walk-to-work accessibility assessment for floating offshore wind turbines. Ocean Engineering, 116:216–225, 2016.

16. M. Martini, A. Jurado, R. Guanche, and I. Losada. Probabilistic assessment of floating wind turbine access by catamaran vessel. Energy Procedia, 94:249–260, 2016.

17. Windcat MK4 Windcat 101 http://www.windcatworkboats.com/portfolio-items/windcat-mk4/ [Date accessed: 2017-02-27].

18. Experimental Comparison of Dynamic Responses of a Tension Moored Floating Wind Turbine Platform with and without Spring Dampers (2015) Journal of Physics: Conference Series, Volume 628, conference 1, Presented at DAMAS 2015, Ghent, Belgium, doi:10.1088/1742-6596/628/1/012056, C. Wright, K. O'Sullivan, J. Murphy, V. Pakrashi

19. Jonkman J. et al. 2009. Definition of a 5-MW Reference Wind Turbine for Offshore System Development.Technical Report NREL/TP-500–38060.

20. DNV-GL. 2007. Environmental conditions and environmental loads. Recommended Practice DNV-RP-C20.

21. Ramachandran, G.K.V., Robertson, A., Jonkman, J.M. and Masciola, M.D., 2013, June. Investigation of response amplitude operators for floating offshore wind turbines. In *The Twenty-third International Offshore and Polar Engineering Conference*. International Society of Offshore and Polar Engineers.

8. AUTHORS BIOGRAPHY

Matthew Shanley is a Ph.D. Candidate in Engineering at MaREI Research in UCC, the title of his work is 'Innovative Hull Design for Servicing Offshore Wind Turbines'. Matthew is also the Quality Systems/Testing Engineer in the Lir National Ocean Test Facility at the MaREI centre in UCC.

Chris Wright is a Ph.D. Candidate in Engineering at MaREI Research in UCC, the title of his work is 'Design Uncertainties for Tension Moored Wind Turbines'.

Cian Desmond holds a Master's degree in Renewable Energy Systems Technology and a PhD in Wind Energy. His current work is focused on the physical and numerical analysis of floating wind energy technologies.

Aldert Otter received a Master's degree in Marine Renewable Energy from University College Cork, after finishing a 15-year long career at sea. Aldert is currently working as a researcher at the MaREI centre in UCC.

Jimmy Murphy is the Lir General Manager at MaREI centre in UCC.

UNDERSTANDING THE TRANSFER PERFORMANCE OF FAST CTV

S Phillips, H Maclean, C Armstrong & I Shin, Seaspeed Marine Consulting Ltd, UK

SUMMARY

This paper summarises the findings of recent research into the transfer performance of fast crew transfer vessels when pushing-on to wind turbine tower docking poles in rough weather.

This research was largely funded by The Carbon Trust and undertaken by Seaspeed Marine Consulting over the past 24 months using physical model testing and full scale sea trials in order to understand and quantify the transfer performance of these craft.

The paper outlines the vessel design parameters that affect transfer performance and describes the mechanisms that lead to fender slip and other important transfer limiting issues.

Typical CTV push-on transfer performance is quantified in the form of performance plots (P-Plots) and the issue of safe transfer limits is discussed.

1. INTRODUCTION

The offshore wind sector traditionally uses small fast crew transfer vessels (CTV) to provide servicing and maintenance resources to the wind turbines. The payload required to complete these roles is generally small, meaning that the vessel size is determined less by payload requirement and more by the capability required to operate the vessel to, from and around the wind farm in the expected sea conditions.

In general, the vessels' capability to perform the required tasks is limited in either transit or transfer mode by their ability to handle the sea-states encountered.

In this paper the term transfer mode is relevant to the push-on step-across transfer of technicians to and from the wind turbine. The performance of the CTV to transfer is a measurement of the vessels ability to allow safe working conditions on the bow of the vessel and a predictable stable platform giving the technicians the confidence to safely step across on to the ladder of the wind turbine.

This specific research was aimed at providing a better understanding of the transfer performance of these craft and, in particular, the typical fender slip and associated failure mechanisms that limit transfer performance.

The research was based on the results from a large number of scale model test experiments and full scale trials, all investigating transfer performance.

The overall research programme has been described in previous RINA technical papers (Refs 1 and 2). For further background information reference should be made to these papers.

2. GENERIC VESSEL DESIGNS

For the scale model experiments, four generic hull forms were designed based on typical CTVs in the offshore wind sector. Although catamaran vessels of different shapes, sizes and propulsion arrangements dominate this sector, this research also included the testing of monohull and SWATH hull forms. The hull forms investigated comprised a monohull, SWATH and catamaran, all propeller powered, along with a waterjet powered catamaran (Fig 1).

Figure 1: Free running model testing – Catamaran, Monohull and SWATH pushed on to model turbine tower

The hull forms were designed to create a family of hulls while incorporating typical hull shapes found in the industry. The overriding parameters such as displacement, waterline length and freeboard were kept constant across all models.

Physical models were built and tested at 1/10th scale for a 20 metre waterline length and 65 tonne displacement.

The bollard thrust force at 85% MCR was taken as 110 kN. The test results were also scaled at 8th and 12th scale (accounting also for displacement and bollard thrust variations) to give results for 16 metre and 24 metre waterlines lengths. This range thus covered typical craft from 18 metres to 26 metres overall length.

Instrumented full scale tests were also conducted, including discussions with vessel operators and skippers in order to determine typical acceptability limits.

3. TEST ARRANGEMENTS

To assess the vessel transfer performance and to understand the associated fender slip mechanisms, free running models were tested, pushing onto docking poles, using two different fender attachments.

Fixed fender – The fixed fender tests used an instrumented vessel fender arrangement attached to a docking pole using a flexible coupling. This allowed the model to roll, yaw and pitch about the flexible coupling but always stay attached to the connection point without slipping, as shown in Figure 2. The fender was coupled to the model using a load cell arrangement such that the longitudinal and vertical forces on the fender were measured and recorded. Using a local motion reference unit (MRU), these forces were analysed to give data relevant to the plane of the docking poles.

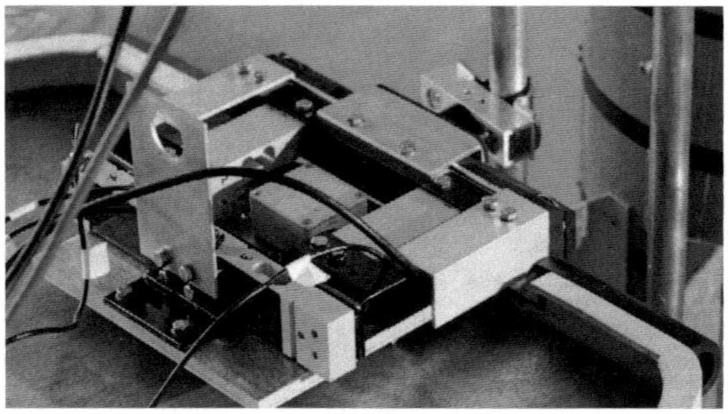

Figure 2: Fixed fender arrangement

Free fender – The second method was to use the same fender arrangement with the load cells but with a representative rubber section on the front mimicking fenders used in the industry, which could be pushed on to the docking poles to allow a more realistic push on trial to be monitored. This arrangement allowed the vessel to slip and/or come away from the docking poles. Figure 3 shows the free fender arrangement.

Both fender arrangements were tested over a range of seastates, varying wave height and period, and across the range of hull forms and propulsion setups for each of the five major transfer headings. The seastates were defined as JONSWAP irregular sea spectra and ranged in significant wave height from 1.0 to 2.5 metres with zero-crossing periods from 4 to 6 seconds. The test set-up was arranged such that the boat landing was orientated for the vessel to push on in head, bow quartering, beam, stern quartering and following seas.

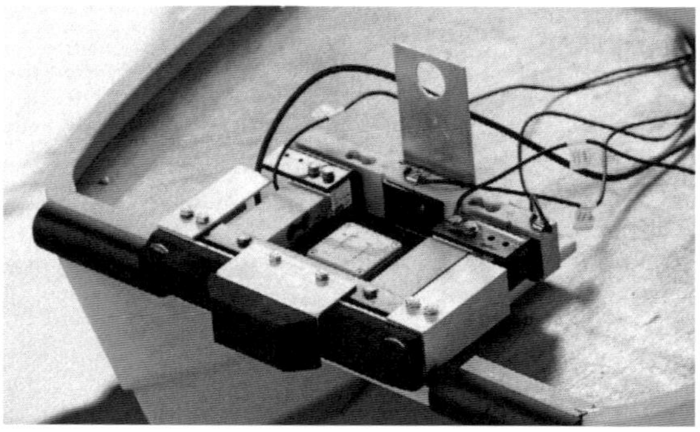

Figure 3: Free fender arrangement

Full scale results were collected using the Seaspeed VMS (vessel monitoring system), initially just monitoring motions and accelerations but later monitoring fender forces using a load cell mounted fender.

4. SLIP MECHANISMS

The mechanisms that caused the slips were largely consistent across the hull forms, sea states and periods but depended on the vessel heading to the oncoming waves. To show this, Figures 4 to 6 represent a time sequence of slips occurring in head, beam and stern seas.

Each figure shows a time series across the period of the slip occurrence. The wave amplitude is measured by a wave probe mounted in-line with the tower to measure the waves as they pass the docking poles. The force measurement is taken on the vessels foredeck and then corrected for angular orientation to give the forces parallel and perpendicular to the docking bars. The frictional coefficient is calculated from the fender force data (vertical force / net horizontal thrust force). The roll and pitch measurement provide for an understanding of the vessels movements preceding and during a slip. The shaft torque is measured to show the occasions that the propellers ventilate – this is shown by a drop in torque as can be seen in Figure 6.

The diagram at the bottom of each figure is a plot of the longitudinal force against the vertical force. By plotting the constant frictional coefficient (Cf) lines, the plots illustrate that the slips occur at high frictional coefficients which are not limited to the highest peaks in vertical force. The plots show data for a 10 minute run.

Vertical force is measured on the fender such that a slip up produces a negative vertical force on the fender. The frictional coefficient of the rubber fender on the docking pole is about 0.8 to 1.0. Coefficients calculated to be greater than about 1.0 are due to the net thrust force approaching zero and hence are largely meaningless.

4.1 HEAD SEA

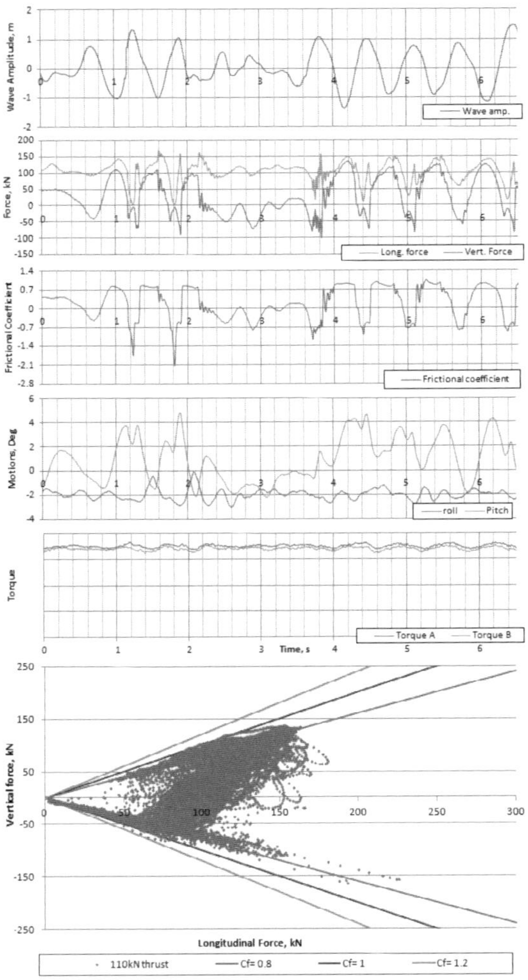

Figure 4: Slip mechanism in a Head sea

The plot in Figure 4 shows the forces and motions recorded during a series of fender slips. At about 3.7 seconds (model scale time) the ship's fender vertical force is about -75kN (the wave pushing vertically up on the vessel) and the net thrust force is about 95kN: this gives a friction coefficient of about 0.8 and so the fender slips up until about 3.8 seconds when it stops. At 4.0 seconds the wave has passed and gravity is pulling the bow down, with a vertical force of about 80kN at which point the thrust force has increased to only 100kN, again giving a frictional coefficient of 0.8 and so the bow slips down until about 4.25 seconds when it grips again.

The trend in the head sea slips was to initially slip up and then back down on the next large wave trough. The initial slips occur at low longitudinal force values when the wave force largely overcomes the push on thrust.

4.2 BEAM SEA

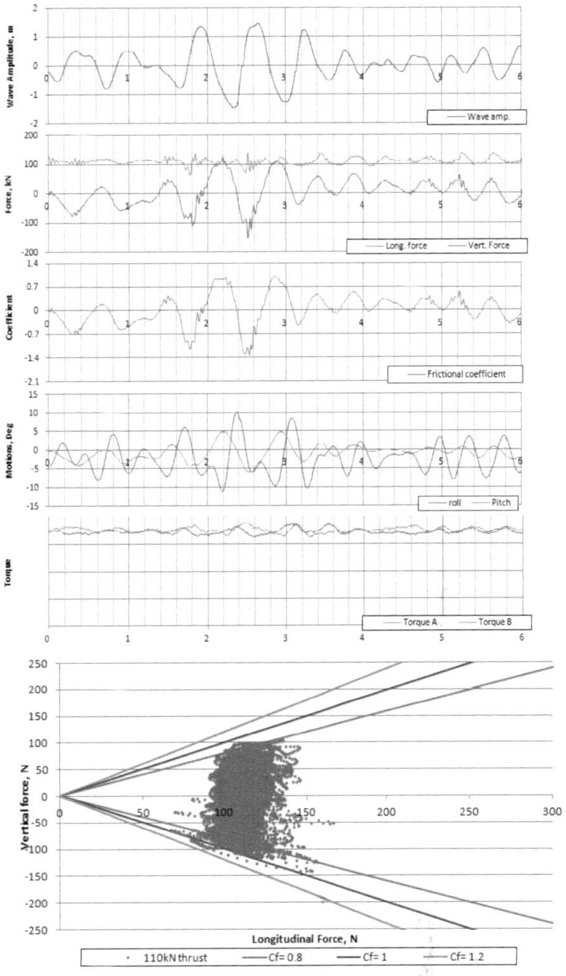

Figure 5: Slip mechanism in a Beam sea

The force plot in Figure 5 illustrates the reduced variation in net thrust (longitudinal) force in a beam sea compared to the equivalent head sea plot in Figure 4. This is also reflected in the time series comparison between Figures 4 and 5. The more consistent longitudinal force leads to the slips only happening during large vertical force occurrences as can be seen in the time series (see slip between 2.0 and 2.2 seconds). The roll variations seen in beam seas are indicative of the fender only keeping in contact with one of the docking bars as it yawed in the beam waves. This allowed the vessel to 'walk' up and down the docking bars reducing the build up of vertical force when a series of large waves passed the vessel. This mechanism reduced the number of slips in the three headings where the waves were beam to the vessel (bow quartering, beam and stern quartering seas).

4.3 STERN SEA

The stern sea plots show the largest variation in longitudinal force and it is no coincidence that this heading also experienced the largest slip events. In the time series it can be seen that the shaft torque drops on two occasions; these are propeller ventilations caused by the propulsion systems proximity to the surface. These cause large slip events as the vessels comes away from the docking poles and then accelerates back towards it creating very high longitudinal impacts, up to and indeed over 400 kN of force.

5. TRANSFER PERFORMANCE PLOT (P-PLOT)

One of the aims of this research was to generate a performance plot which allows the transfer capability of a vessel to be easily portrayed and to be used as a baseline for comparisons to current industry standard performance. The radial plot in Figure 7 shows the limit lines derived from this work estimating the sea state for each of the major headings that the vessel will be able to safely transfer. The limiting criteria used for this plot is a 95% confidence of the vessel not slipping. That is equivalent to 1 slip in 20 waves.

The limits are broken down into three categories, friction, motion (roll) and freeboard. The friction limit is the derived limit from the model tests defining the maximum significant wave height at which the vessel will slip no more than 1 in 20 waves. The roll limit is the wave height in which the vessel will exceed a roll rms of 3 degrees, this being considered to represent the threshold level of motion for safe working on the foredeck. Roll motion is dependent on the wave period and so short, standard and long periods are represented on the plot.

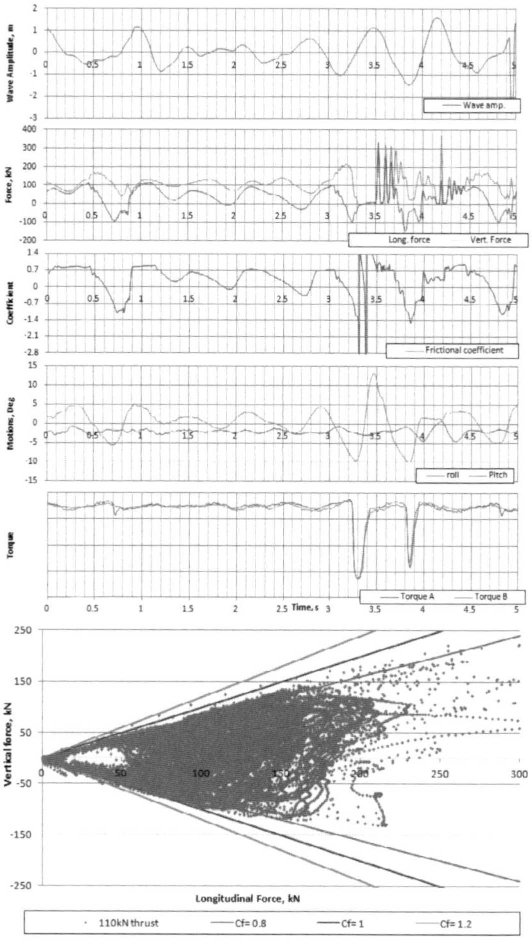

Figure 6: Slip mechanism in a Stern sea

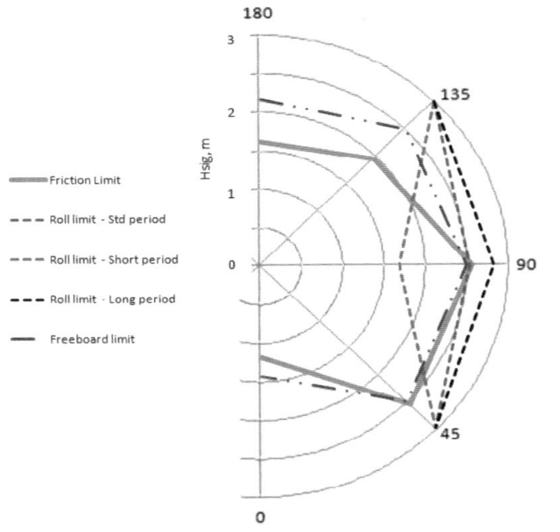

Figure 7: Transfer P-Plot, 20m Lwl Catamaran (prop)

The freeboard limit is the height at which the waves reach the wet deck of the vessel or come over the bulwark. This limit occurs at a set height above which the vessel will not be able to transfer.

Interestingly, the performance and limiting sea conditions for the monohull and catamaran were very similar (see Fig 7 & 8). The SWATH, on the other hand, showed a significant benefit in head and bow seas (see Fig 9) due to the slender hulls and hence low vertical wave induced buoyancy forces. However, due to the exposed propellers and low heave damping, propeller ventilation and consequent loss of thrust, significantly curtailed this benefit in stern seas.

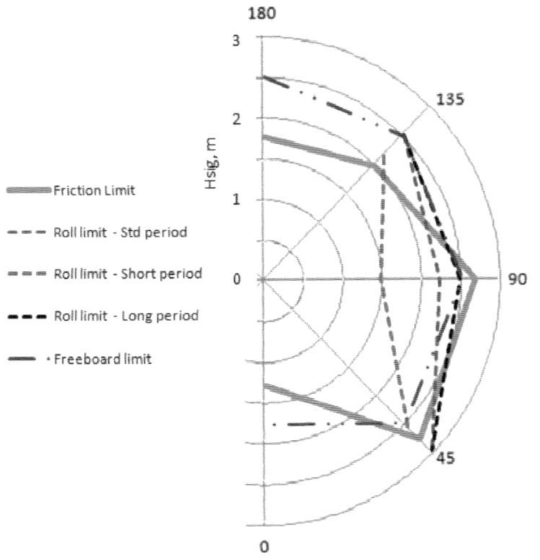

Figure 8: Transfer P-Plot, 20m Lwl Monohull (prop)

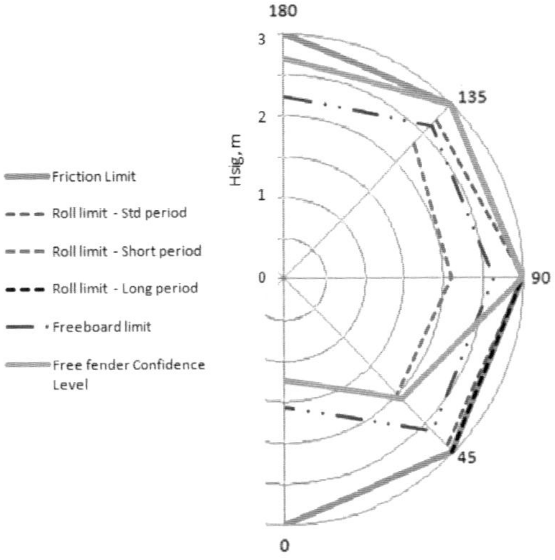

Figure 9: Transfer P-Plot, 20m Lwl SWATH (prop)

6. VESSEL DESIGN ISSUES

6.1 TRANSFER FAILURE MECHANISMS

There appear to be a number of mechanisms associated with transfer failure – primarily fender slip but also loss of fender contact, water on deck and deck motion. An understanding of their nature is considered to be a pre-requisite to improved vessel design.

Fender slip occurs when the friction between the docking pole and the fender surface is overcome – this can be due to reduced thrust (propulsion failure or ventilation of propulsor), excess buoyancy force due to wave amplitude (or conversely loss of buoyancy due to a wave trough, particularly if the bow has 'climbed' the docking poles), excess vertical force due to contact of wave with the wet deck of a multihull, excess longitudinal force acting against the bollard thrust due to wave slope and/or wave orbital motion and, occasionally, excess lateral force due to current (such as tide). Related occurrences of slip also occur if the fender material starts to break down or the coefficient of friction is dramatically reduced due to wetness or excessive slime build-up.

In some instances, particularly with very fine bowed vessels such as SWATH craft, the longitudinal force of the wave (due to slope and/or orbital motion) can become larger than the bollard thrust before a frictional slip occurs and in that instance the fender can come away from the docking poles without actually slipping (then generally impacting back on, hard, due to the wave induced longitudinal force changing phase and reinforcing the bollard thrust).

Another limitation that has been observed, mainly in large beam seas, is the significant swell-up of a wave impacting the side of the vessel bow and breaking over the foredeck and transfer point.

Finally, an excess of motion on the foredeck can lead to it being insufficiently safe for personnel to walk to the transfer point. This is normally associated with rms roll angles greater than about 3 degrees.

6.2 DESIGNING FOR EXTENDED TRANSFER CAPABILITY

For conventional push-on fender transfer, the primary design issues are:

a. The level of bollard thrust – in general the higher the thrust the greater the vessel capability (see Fig 10 below).
b. The freeboard of the vessel, particularly that of the wet-deck (mainly forward but also aft, the latter in the case of stern seas) – the higher the freeboard the greater the vessel capability (up to a bow freeboard probably not exceeding about 18% of the waterline length)
c. The coefficient of friction – this is largely fixed by the material properties.
d. The susceptibility of the propulsor to ventilation during reduced immersion at the stern – in general the propulsors should be as low and as far forward as is commensurate with good propulsive and manoeuvring efficiency.
e. The reserve buoyancy in the bow area of the vessel – in general a relatively fine bow can be beneficial, although issues such as transit and approach performance can be adversely affected by making a very fine bow.
f. The zero speed roll motion of the vessel – this should be as highly damped as is feasible.

All these issues can be addressed to provide for an extended transfer capability – and although it is acknowledged that some changes may have a negative impact on transit or payload carrying capabilities, some will also improve the other modes of operation.

However it is very important to note that in extending the safe transfer limit to higher sea states, this can lead to less safe vessel motions once there is a failure.

Even if that failure only occurs very infrequently, it recommended that its effects be accounted for in the design, layout and operation of the vessel.

The current market analysis indicates that the smaller vessels generally have greater bollard thrust compared to their displacement than the larger vessels, which leads to the underlying limit of these vessel being related to freeboard and wet deck height rather than thrust. The larger vessels generally have lower bollard thrust for their displacement, thus limiting their push-on performance due to limited thrust. The 20m waterline vessel size is the most common in the current market and, at present, seems to have the most appropriate balance between these transfer parameters.

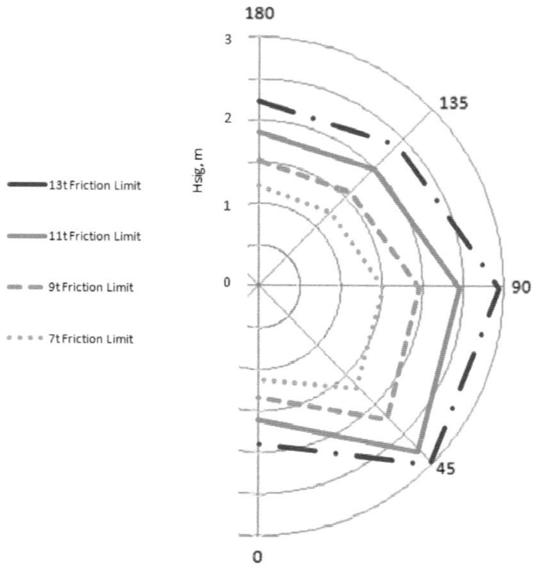

Figure 10: P-Plot showing effect of push on thrust for 20m Lwl Catamaran (prop)

7. SAFE TRANSFER LIMITS

The transfer limits presented in this paper are largely based on the statistical likelihood of experiencing a slip no more than once in 20 waves – or approximately once every two minutes, or to put it another way, a 95% confidence limit of no-slip. This is based on our notes from sea trials but it is acknowledged that this is a contentious issue and still needs to be finally determined and agreed within the industry. Another related issue is the size of an acceptable or unacceptable slip (maximum acceptable slip values of 0.3 to 0.5 metres have been suggested).

Other parameters include an rms roll angle of greater than 3 degrees, which can lead to difficult lateral motions on the foredeck, and water over the deck (which is generally concurrent with a slip or near-slip occurrence).

From experience of sea trials and from analysing the physical model test data, it has been concluded that the measurement of fender forces (net longitudinal thrust and vertical forces) is the only reliable method for assessing the likelihood of a fender slip or detachment occurring. It is thus expected that fender force load measurements will become a more common and important feature for the maintenance of transfer safety and for the monitoring of decisions to undertake or abandon transfer operations.

Monitoring of fender forces also raises the prospect of being able to predict when slips might occur, thus providing a real-time warning system. Such systems are currently in the R&D stage.

8. CONCLUSIONS

This paper presents the findings of the research into the transfer performance of fast Crew Transfer Vessels being used in the offshore wind sector. It discusses the performance, limitations and underlying mechanisms involved in this important mode of operation.

It is intended that this research be reinforced using further data collected from operational vessels using the vessel performance monitoring systems, to further validate and improve the performance plots. These can then be used by the offshore wind service providers to better understand the limitations of their vessels and to more accurately assess the commercial implications of

9. REFERENCES

1. S.Phillips et al, 'Crew Transfer Vessel Performance Evaluation, RINA, 2015.

2. H.Maclean et al, 'Fast Crew Transfer Vessels – Transit and Transfer Performance Research, RINA, 2016.

10. AUTHORS' BIOGRAPHY

Stephen Phillips is the Managing Director at Seaspeed. He was previously Chief Naval Architect (Patrol Craft) at VT Shipbuilding (UK) and Propulsion Systems Manager at Incat Designs Pty Ltd in Australia.

Hugh Maclean is a Project Manager at Seaspeed and is responsible for specific company projects. He is a Naval Architect graduate from Southampton University with experience in hydrodynamics and autonomous marine systems.

Charles Armstrong is the Trials Manager at Seaspeed and is responsible for the undertaking of sea trials and model tests. He is a Naval Architect graduate from Newcastle University.

Iebum Shin is the Design Manager at Seaspeed and is responsible for the running of the company's technical office. He is a mechanical engineering graduate from Hongik University in South Korea and a Ship Science graduate from Southampton University.

SHIP DESIGN AND VALIDATION BY USING HARDWARE IN THE LOOP SIMULATIONS

J.C. Kalis MSc and **S. Ivancsics MSc**, Damen Shipyards, NL

SUMMARY

Conventional testing procedures focus only on the technical performance of the items that are covered in the shipbuilding contract and do not pay proper attention to the real life conditions that the vessel has to operate in. This may lead to sub optimal operational performance of complex vessels. This paper introduces the combined use of hardware in the loop testing (HIL), generic numerical modelling and physical model testing for the verification and validation of the design and technical performance, in real operational conditions, of a Walk2Work vessel. Arguably this is the only viable method to assess and optimise the complex interaction between the vessel, its systems and the harsh environment in which it operates.

1. INTRODUCTION

Conventional procedures for specifying the performance of both gangway systems and DP systems fail to address the typical intricacies of this specific operation; DP performance assessments are generally done on the basis of so called DP plots which base the DP performance on a static balance between the mean environmental forces and the thrust that can be generated by the DP systems thrusters and the gangway systems are characterised against the first order or wave frequent vessel motions. Both approaches fail to address the interdependency between the systems and the different timescales on which each of the systems work.

1.1 STATIC AND DYNAMIC DP ANALYSIS

The primary assumption that is made for doing a static DP analysis is that once the vessel is able to withstand the mean environmental forces it is capable of staying in position. To account for dynamic effects a thrust degradation factor is usually taken into account to compensate for the effect that on average the thrusters will not provide the full power due to the oscillatory nature of the DP operation as indicated in below graph (Fig 1).

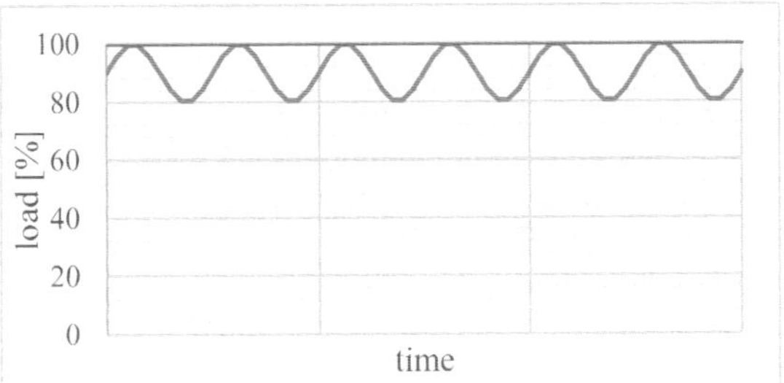

Fig 1; schematic graph of power requirement and available power in DP

The result of this analysis is still binary: the vessel will stay in position or it won't.

A static analysis fails to address that a DP system is a reactive control system and therefore it will never exactly be in the required position; it constantly tries to minimise a positional error but it is only reacting on a displacement that has taken place already. Furthermore a DP system will only compensate low frequent displacements and no wave frequent displacements. The result is a vessel that continuously moves around the predefined DP setpoint and depending on the operation the <u>residual motion</u> determines whether or not an operation can be carried out.

A dynamic DP analysis is addressing the fact that the residual motions are in reality determining the feasibility of an operation and not the mean environmental forces.
In the example that is used in the next paragraph an environmental condition is used that in a static DP analysis resulted in a thrust utilisation of approx. 70% of the available thrust when all the forces act perpendicular on the vessel. The initial dynamic DP analysis based on a generic DP controller resulted in a lateral position standard deviation or root mean square value of 2.7 m.

1.2 GANGWAY ANALYSIS INCL. DP ACCURACY

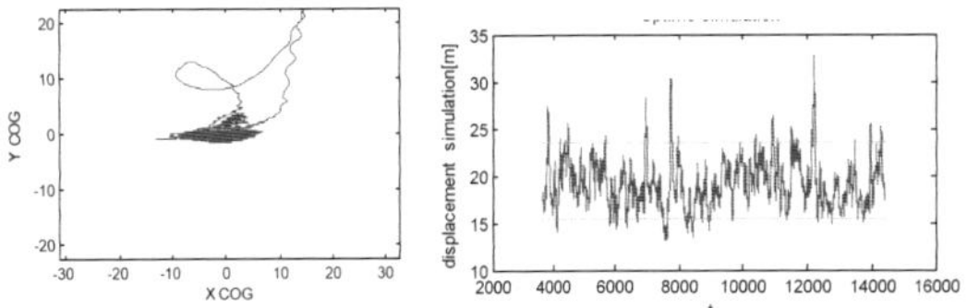

Fig 2; time trace of vessel position left and required compensation distance of the gangway right from time domain DP accuracy simulations.

In the above two graphs (Fig 2) a single result of the initial time domain analysis of the DP accuracy of the vessel (left figure) in relation to the distance that the gangway needs to compensate (right figure). In this example the wave frequent vessel and low frequent residual DP motion each individually were not large enough to exceed the distance limits of the access system, but in the right graph it is clear that occasionally the required compensation distance (blue oscillatory line) exceeds the limits (red horizontal lines) of the gangway system under consideration.

It needs no further explanation that the risk of an access system being pushed outside it's working envelope needs to be reduced to acceptable levels and from above example it is clear that a proper assessment can only be done on basis of an analysis that includes both high and low frequency effects of a vessel in a seaway controlled by a DP system.

1.3 INTRODUCTION OF CASE STUDY

In this paper the proposed testing procedure is explained using the Dame Walk2Work vessel design as case study. As explained in earlier papers and presentations on the Damen Offshore wind Walk2Work design a novel design strategy was adopted to be able to design a revolutionary vessel rather than offering a modified existing design. The benefits for this choice were discussed in detail in [Kalis, 2014] and focussed on the ability to match the vessel design and it's systems to the final mission and functional requirements.

1.3(a) Systems Based Design Approach

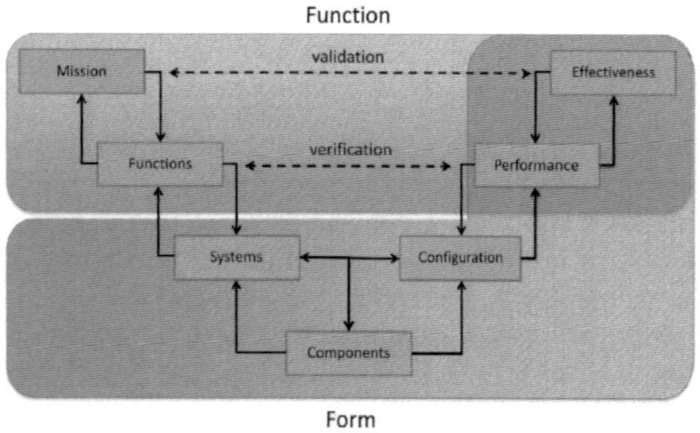

Fig 3; [DENUCHI] System Based Design (SBD). Framework illustrating the relationship between function and form

The prime reasons for selecting a systems based design methodology for designing this vessel was the assumption that for a revolutionary design for a completely new set of functionalities the traditional way of modifying an existing design is leading to an inefficient design. Based on a complete understanding of the mission of the vessel a system based engineering approach will lead to a vessel design that complies with all the functional requirements for fulfilling the mission without inheriting design inefficiencies from earlier design challenges. It also allows the designer to balance the functionalities in a completely different way to traditional evolutionary vessel designs, something that is considered essential for an immature market as the offshore wind industry still is.

As is stressed in the work of [Denucci, 2012] It is extremely important to quantify design assumptions instead of accepting conventional wisdom; this is especially true for immature markets where standard procedures and common wisdom are no relevant source of information. All that is available is at best from adjacent market segments and not dedicated for the specific niche for which this design had been made.

The process of quantifying the design assumptions is started relatively early in the process by defining the mission of the vessel in measurable quantities or capacities of systems. This is done on the basis of a very rough functional decomposition for the functions that are required for fulfilling the vessel mission as proposed in [Wolff, 2000]. In figure 4 an example is given for some of the systems and its capacities with fictional numbers.

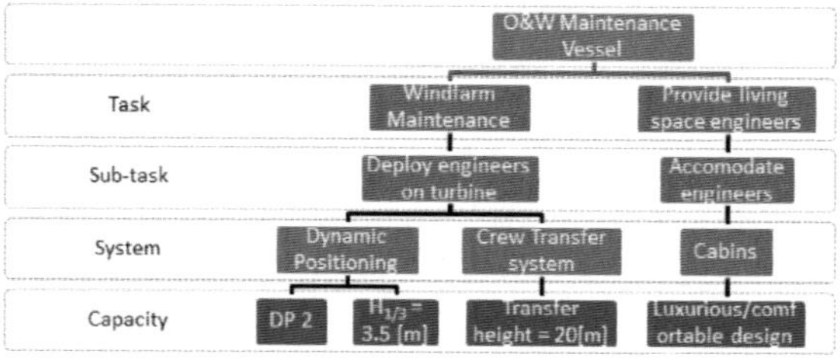

Fig 4; Functional decomposition.

Based on the interviews that were held with end-users and possible operators of the vessel the primary mission was defined as the capacity to transfer a certain amount of passengers in a limited time to multiple turbines. For the economic viability of the concept this task needs to be able to be carried out in a minimum amount of days per year in a safe way.

1.3(b) Verification and Validation

In order to avoid further confusion on the concepts of validation and verification a short side step is made to define the difference in the view of the authors based on a citation of the definition from Wikipedia.

Validation: The assurance that a product, service, or system meets the needs of the customer and other identified stakeholders.

Verification: The evaluation of whether or not a product, service, or system complies with a regulation, requirement, specification, or imposed condition.

These definitions are in line with the definition as used in the work of [Denucci, 2012] that was also used in the earlier paper on this vessel design and they clearly make a distinction between evaluating requirements for individual systems and assessing the effectiveness in fulfilling the mission of a vessel.

1.4 VERIFICATION AND SYSTEM DESIGN

A very important benefit of the system based design approach is that it creates early on an awareness on the necessity to quantify separate design aspects, since per system that is being designed the prime system capacity in relation to its functional requirement is explicitly written down. In this paragraph two design items will be discussed that were tested later in the design verification and validation process.

1.4(a) Main Dimensions and Seakeeping Behaviour

During the very first design steps a variation of main parameters was carried out to check whether the defined design space was sufficiently large to create significant differences in seakeeping behaviour. The definition of the design space may have resulted in a sub-optimal solution for sea keeping behaviour of the vessel but it was considered necessary since primarily the stability requirements for a 60+ person passenger vessel are fairly complex. The design space was chosen such that the designers were fairly certain that they would be able to meet the stability requirements and from the main dimension analysis it turned out that the seakeeping behaviour could indeed be influenced significantly. In the two diagrams in figure 3 the relative change in lateral and vertical accelerations at the central technician accommodation are presented for 5 different hull shapes and corresponding loading conditions.

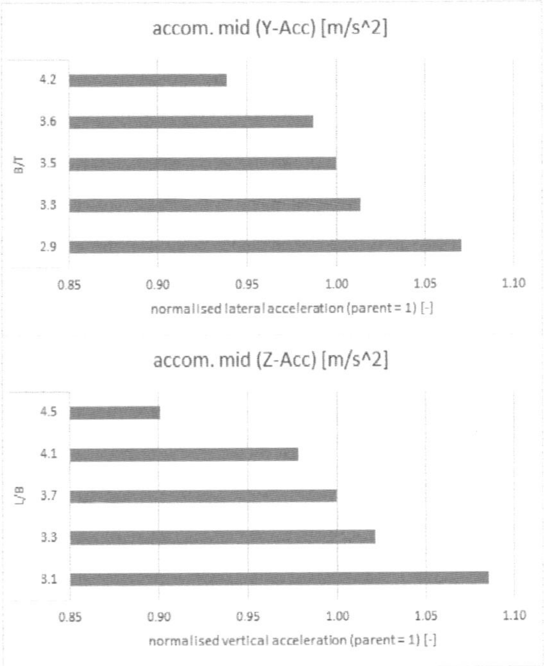

Fig 5; lateral acceleration (Y-Acc) and vertical acceleration (Z-Acc) in central technician accommodation for 5 different hull shapes

Considering the above set of graphs it will be no surprise that the final vessel dimensions are B/T = 4.2 and /B = 4.5. The very low accelerations were confirmed in the model tests where in 2.5 m. beam seas and 30 kn. wind the roll angle was only 1° RMS.

1.4(b) Motion Compensated Gangway Selection

Based on the initial DP accuracy simulations all at that moment commercially available motion compensated gangways were compared against the expected total high and low frequency vessel motions. The range of systems varied from 3 degrees of freedom to 9 degree of freedom systems. Based on this assessment a 3 degree of freedom system was selected and at the moment of writing this paper the gangway control system is being tested in conjunction with the DP control system in a HIL setting as explained later in the paper.

1.5 VALIDATION AND VESSEL DESIGN

Early in the design process a discussion was started with the final owner of the vessel on how to ascertain the annual workability of the vessel and the systems supporting the prime mission of the vessel. This revolved around the operational performance of the combination of primarily the DP system in combination with the capacities of the gangway system.

The interaction between the different systems are schematically drawn in figure 3 and are used to design the validation programme for this vessel design.

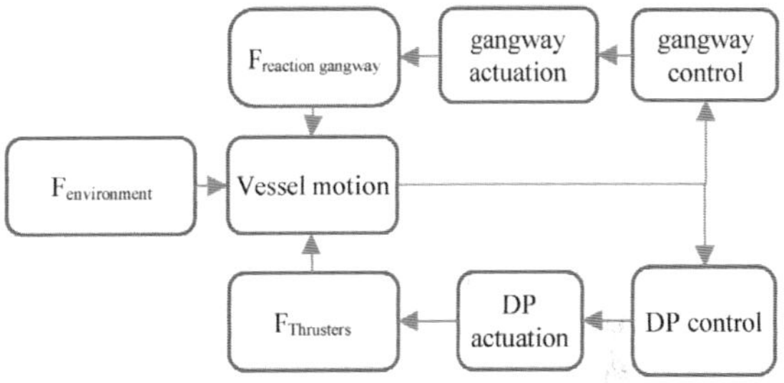

Fig 6; Dependencies between vessel motion, DP and gangway controller

Since both the DP and the access system are dependent on the behaviour of the vessel and each other it is not possible to assess the overall vessel performance for transferring technicians by separately considering these systems. How this is tested is described in the following chapter and at the end of the paper the practical aspect, i.e. is it possible to safely transfer people to a turbine is discussed shortly at the end of the paper.

2. INTRODUCTION HIL TESTING

The high complexity of the automation systems on technologically advanced vessels is creating the need for an integrated approach for system development and optimization. Real time testing of the control systems on board is expensive and time consuming. Modelling and simulation is an effective method for the developer to create a controlled virtual environment comprising of the subsystems of the vessel. HIL simulations allow testing and tuning of selected control systems before applying in the real vessel where only fine tuning is taking place resulting in short commissioning and sea trial time.

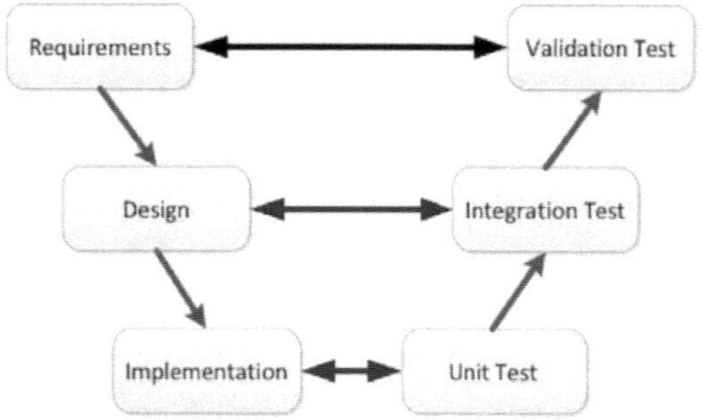

Fig 7; Product development V curve

2.1 THE HIL CONCEPT

The general terminology of HIL refers to the phenomenon when a physical part of the system is replaced by simulation. These simulations provide numerous opportunities for industrial partners to improve the integrated product by collaboration. First of all, during the testing process the implementation and unit test can be skipped. Furthermore, this is an opportunity to prove a feasible system design.

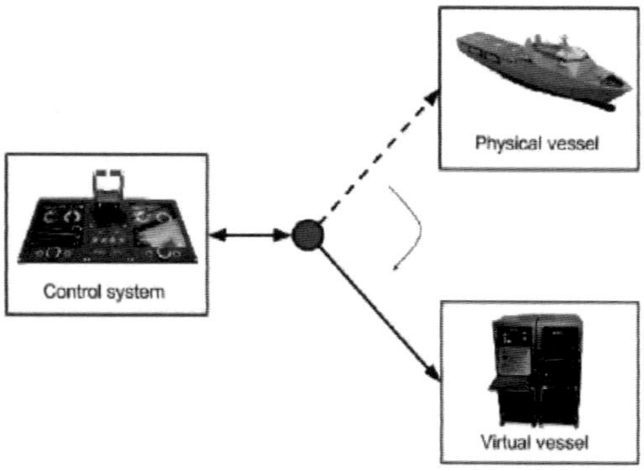

Fig 8; Hardware in the loop operational concept

2.1 (a) Scope

The scope of the HIL simulation is tailored to its specific target, and usually based on the following criteria:

- Rules and regulations: class, regulations in the environment the vessel is used (country, international waters)
- Functional specification and requirements
- User philosophy
- User manual

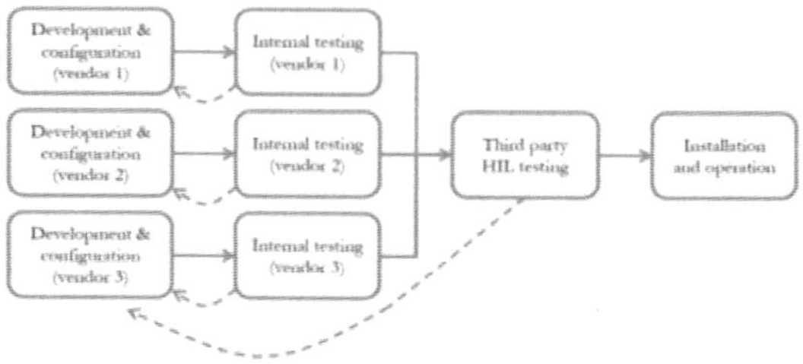

Fig 9; Enhanced procurement procedure

The HIL test program therefore consist of multiple type of tests:

- Functional testing: Verification of the system functionality and components
- Failure mode analysis (FMEA) : Testing of failure detection and handling
- Integration: Testing the integration physically and conceptually between the components
- Performance testing: The key naturally of a HIL test, if the given system fulfils its requirement specification under different operational and environmental conditions. It requires high fidelity models and subjected to careful analysis of model accuracy and sensitivity.

2.2 MOTIVATION

The shipping industry is rather conservative, which means that its building blocks mostly consist of proven designs. Since failures can be extremely dangerous or costly, the idea of HIL was introduced to overcome the over constrained testing and look into possibilities which would not be possible to test in real environment. The benefits of performing HIL simulations are categorized as follows:

- Help commissioning: Checking if the pre-defined settings has to be modified. Calibration and interfacing check is already done in an early phase of the project.
- Checking functionality: fundamental behaviour of the control system is correct. Investigates if all interlock during an operation is present. By interaction of subsystem and high fidelity models a good estimation of initial parameter settings before the trial can be determined.
- Reduced risk: In case of troubleshooting, the scenario where the mistake occurred is easily reproduced, since there are circumstances which occur seldom in specific areas of the world, e.g., freezing cold in Vietnam.

2.3 EXECUTION

The idea of replacing part of an application with a simulator is simple, albeit during the implementation one would face difficulties. There are some challenges which has to be overcome in the simulation phase. If a digital communication interface is not available can lead to a long and tedious wiring process, which hinders the development especially on short deadlines.

2.3 (a) Simulation

Selecting the simulator can be a challenging task since, it has to fulfill many requirements:

- Cost is a driving factor, especially if it is intended to use only in the specific project.
- Real-time: the environment has to be able to perform the simulations in real-time with the necessary sample time in order to guarantee that the outcome approximates well enough the real vessel.
- Modelling: The extent of the modelling is mostly governed by the available information. It is a known fact that often there are not enough information available to have the best fitting models for simulation purposes. There are factors which can make the process easier, if there are not enough available information, such the modeler has already experience with the type of system, so even though there are not enough information available, similar models can be scaled.

2.3 (b) Hardware (in the Loop)

The supplied hardware is developed mostly by a third, external party, whereas the information is-again-restricted. The integrator has to have all the available information or education on the system in order to utilize it to its maximum extent. The sources of the required information are twofold. Firstly, there are the algorithms behind the hardware, how the system principally works what are the key performance indicators and what are the possibilities which the user can change. Secondly, the commissioning knowledge of the system, such how to put it into use, if there are calibration issues how to overcome those.

2.3 (c) Interfacing

Most of the time the scope of HIL testing is either validate functionality or/and improve performance. Therefore, it is not required to hard wire the components, the signal are generated according in the software, therefore, in case of failure in the communication either means faulty algorithms or hardware errors.

2.4 OPPORTUNITIES AND LIMITATIONS

There are tremendous opportunities to utilize HIL simulations. From a low levels subsystem to the abstract control algorithms. During the modelling process the integrator improves his understanding of the components, it becomes easier to make the specifications and requirements.

By using the system, the concept is becoming clear and the user can also understand better the limitations of the system by assessing the available functionality and understanding the uncertainty in the components.

3. HIL

3.1 CONCEPT DESIGN

As it was mentioned the goal of this project was to design, test and validate a novel concept of a gangway system, where the contract not only constrain the integrator (shipbuilder) on the soon to be delivered vessel, but on its performance as well. As it is mentioned the system has to be able to transfer people 95 % of the time in the given worst operational conditions. The integrated schematic system design is depicted in *Fig 10.*

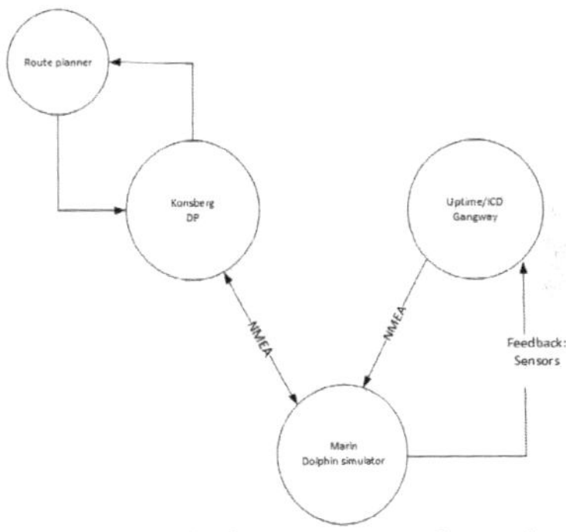

Fig 10; Hardware in the loop system schematic

3.2 MODELLING

As it is visible in *Fig 10* the modelling was performed on their own system by each party. The ship model was developed by MARIN with the help of the Anysim software, based on the model tests results. Furthermore, they have provided the simulation environment to visualize the experiment. (DOPLHIN)

Konsberg were only responsible for the DP system, therefore from their part no additional modelling effort for the HIL was done.

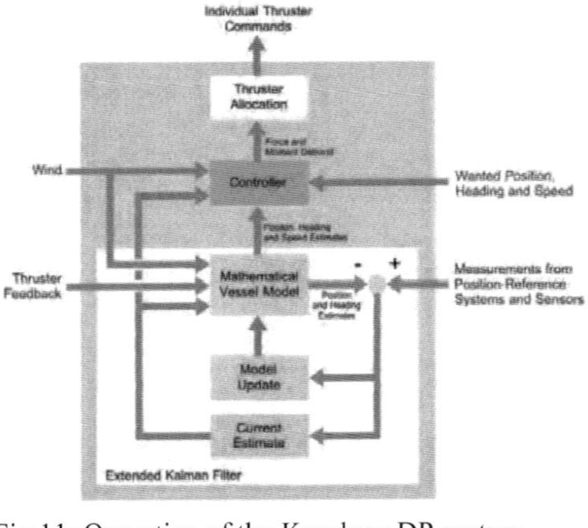

Fig 11; Operation of the Konsberg DP system

The operation of the Konsberg DP system is depicted in Figure-X. It is shown that the control loop consist of multiple subsystems. Based on the thruster feedback, the measurement of the position reference system and the sensors an extended Kalman filter is implemented in order to estimate the states of the vessel. The behaviour of the estimator is depicted in Figure 12.

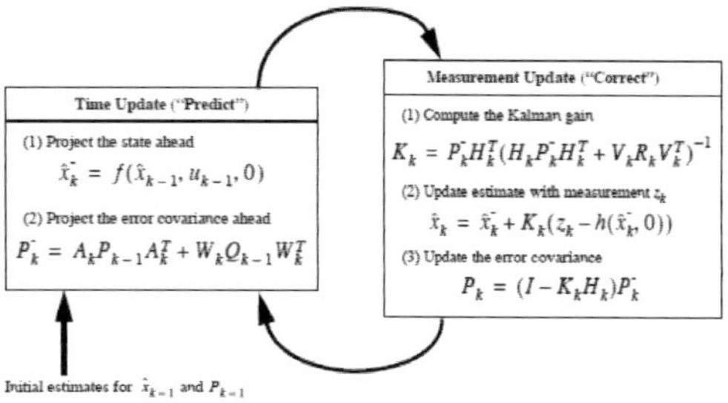

Fig 12; Operation of the extended Kalman filter

The mathematical model of the vessel is running in the simulator, where based on the aforementioned input date the current state (referred as singular since the state is considered a multidimensional vector rather than multiple scalar variables) is predicted. The following step is to calculate the error covariance (P_k). After predicting the states the result has to be corrected by the measurement update. As a first step the Kalman gains are calculated. Based on the gains the state estimation and the error covariance with measurement z_k is updated.
The output of the Kalman filter is connected to the controller which is generating the control signals (force and moment) based on the estimation and the setpoints. The control signal defines the forces and the moments applied on the vessel, but an additional system is required to translate these physical attributes into actual thrust commands to the thrusters.

The third party Uptime with the joint forces of ICD was to develop the gangway model. Since in the simulator the gangway is only visualized they had to adapt their existing simulator to use it in the Real-Time (RT) environment.

3.3 INTERFACING

Interfacing is commonly considered an effortful, albeit a dull part of the project. However, this is a crucial element, where high precision needed, since this is the gateway between the different subsystems. The interfacing can be broken down into an electrical and a physical part.

3.3 (a) Electrical

The subsystems are connected via an electrical platform, where it is ensured that the necessary signals can be communicated. In this project an Ethernet based platform was chosen, therefore connecting the system were rather rudimentary, due to its availability and to the fact that is used by a large group of users for many years now. A necessary step in the process to define the message database, alas what are the message identifiers, how are they built up and their sending frequency.

3.3 (b) System

If the electrical interfacing is done, the parties have to make sure that the signals are communicated in the right way. One of the key differences are that Konsberg, Marin and Gangway (ICD) are using different coordinate systems, therefore the same information is represented in a different manner.

3.4 VALIDATION

Validation is as it was shown is also a multiple phase process. A strategic approach is needed to prove that the performance of the simulator is identical, or at least good enough to approximate the behaviour of the real vessel with a high confidence interval in order to prove the business case.

The actively involved control systems are the DP and the gangway control system. The conditions for a successful operation in the given environment are that the DP system remains within the area of 3 meters and the gangway will be able to compensate the residual motions due to the waves.

3.5 INTEGRATION TESTING

Based on the requirement of the vessel (and client) four scenarios of testing was developed. If the vessel performs in all cases it can be concluded that the conceptual design meets the requirements of the client

3.5 (a) Scenario 1

Scenario 1 is the validation of the model test in which the positioning accuracy of the vessel will be tested in blow off position with respect to the turbine. The conditions are defined as:

Current	$v_{current}$[m/s]	0.77
	Direction [deg]	180
Waves	H_s[m]	2.50
	T_p[s]	7.34
	V[-]	3.3
	Type	JONSWAP
	Direction [deg]	180
Wind	v_{wind}[m/s]	15.43
	Type	Constant
	Direction [deg]	180

3.5 (b) Scenario 2-3

Current	$v_{current}$[m/s]	1.20
	Direction [deg]	180
Waves	H_s[m]	1.40
	T_p[s]	6.05
	V[-]	3.3
	Type	JONSWAP
	Direction [deg]	180
Wind	v_{wind}[m/s]	10.7
	Type	Constant
	Direction [deg]	180

The following approach and abandon situates have to be performed in the conditions mentioned in the aforementioned table.

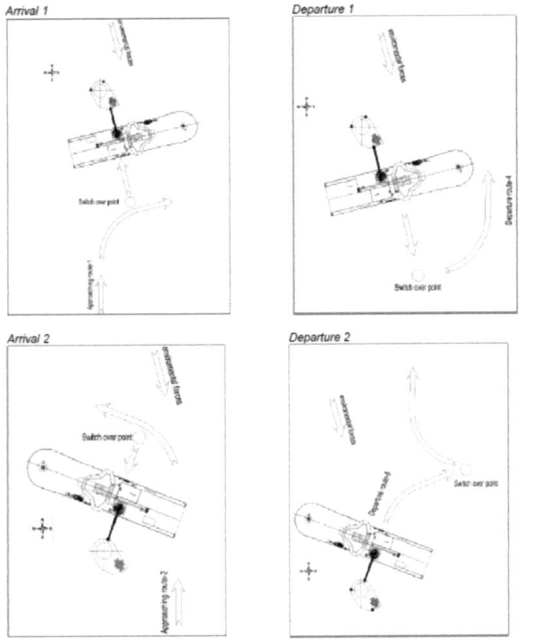

Fig 14; Approach and departure routines tested during physical model testing

All the above procedures have been tested during the physical model testing and are being replicated in the HIL environment. Since the final approach is still controlled by the DP officer the time that is required to accurately position the vessel is dependent on the operator of the DP system. During the physical model tests an experienced DP officer was controlling the vessel whereas the HIL testing is currently only done by an unexperienced engineer. Therefore the results of the physical model tests cannot be compared yet to the HIL test results.

3.5 (c) Scenario 4

In this scenario the sailing capability of the DP system was tested functionally. At the moment of writing the paper no quantitative assessment is available yet. Once these routines are properly tested it is possible to assess the full cycle time from turbine to turbine for each offshore windfarm for which all information regarding the geometry and environmental conditions are available. This is vital input for planning offshore operations for the end user of the vessel before entering into contract.

4. RESULTS

4.1 DP ACCURACY TESTS

As it was described before the operation of the control system is broken down into two component. First, the DP system which makes sure that the vessel remains within a predefined meter circle what is the operational range of the gangway system. In the below graph the vessel footprint is presented for the different methods that were applied for estimating the DP accuracy.

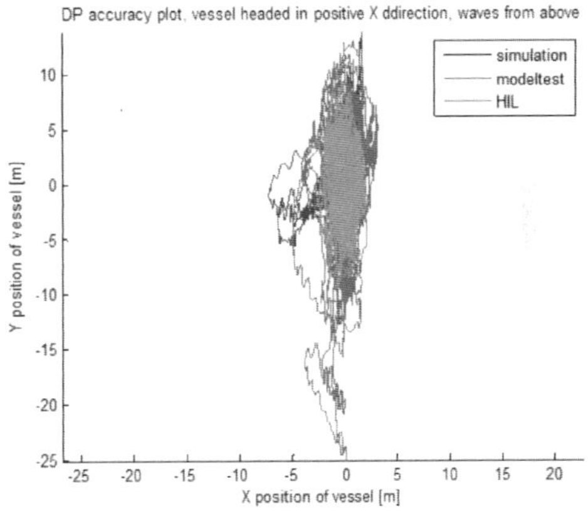

Fig 15; Comparison of the different estimation methods for DP system performance

It is visible that for the period the DP system was operated most of the time the system remained relatively close to the set point over the three hour simulated time. Since from the above plot no probability of a safe operation can be extracted also a more statistical analysis is made of the system performance. Results of the analysis are depicted in Figure 16 for 1 of the most relevant parameters of the system: gangway telescope distance. For all other gangway system parameters similar analyses are made which make it possible to estimate the time that the vessel, DP system and gangway can guarantee a safe transfer of people with a corresponding required statistical confidence.

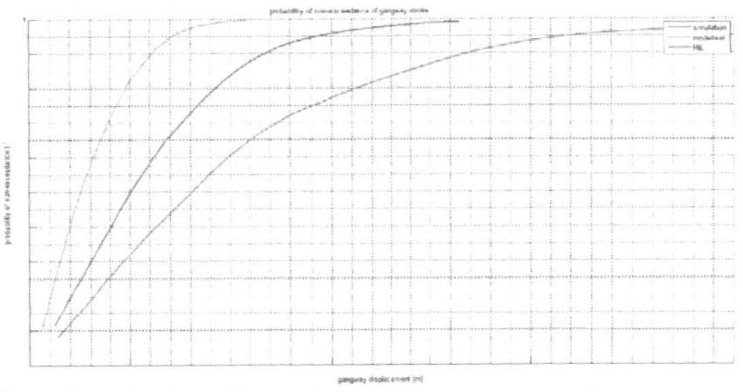

Fig 16; Probability of non-exceedance of gangway telescope stroke

The graph shows the probability for non-exceedance, where the critical distance is on the horizontal axis. The model test results are indicating a significantly lower probability of non-exceedance, however during the HIL and offline simulation probabilities for a successful transfer of people were considerably higher. It is clear that the different methods are resulting in different outcomes. This a primarily caused by the different settings of the DP control system. In the initial offline DP simulations a generic DP control system was applied, in the physical model tests a relatively new DP control system setting with very limited tuning was used. For the presented HIL test results the DP control was further and more rigorously tuned for the specific operation, but further optimisation is still being carried out.

The deviations from the desired position $(x_{set},y_{set})=(0,0)$ is more significant in the Y direction. The reason for that can be found in the design of the control system. As it was mentioned in Section 3-2 that the control system is using Kalman filter to calculate the state of the vessel. In the basic configuration the weight matrices are using the same weighting factor along all axis, however it became clear that each position there is a worst excited direction, therefore the control system was updated. The update was that the gains along the different axis decoupled so the system could be fine-tuned based on the experienced performance.

For the control of the gangway the quantitative assessment is not yet done. In the last few weeks before writing this paper a functional verification has been carried out and the gangway can now be controlled in a reliable and seemingly realistic way. The exact performance of the system can not exactly be given because those scenarios are not yet run.

5. RELEVANCE FOR REAL LIFE OPERATIONS

From the results of the HIL simulator set up it is now possible to define a statistical reliable performance of the combination of the vessel, the DP system and the gangway system. Typically a shipbuilding contract contains individual system performance parameters that are the result of the mission requirement analysis that was made by the operator before contracting a shipyard for the acquisition of the vessel. As is explained in this paper it is not possible to separate each system and individually test its performance and get a reliable outcome for the vessel's capacity to meet the needs of the operator to fulfil the end-user's mission. With the HIL simulations this is possible and at first this leads to more accurate contract conditions but may also enable the shipyard, the operator and contractor to make reliable and realistic performance assessments.

Further the simulator set up enables the crew to train with the vessel in a realistic environment with the identical controls and interfaces as on board of the vessel; this is considered a major benefit in ascertaining safe personnel transfer but for this project the simulator was set up only as an engineering tool. Further developments in the field of training crew will most likely be discussed in the near future with the owner of the first vessel of the series.

6. CONCLUSIONS

A new approach was introduced in the design process of the Walk 2 Work vessel. Due to the strict operational requirement of the vessel a novel so called performance contracting approach was used. The integrator (DAMEN) in order to ensure its client performed an extensive research how to reach the desired goal. The solution was the hardware in the loop simulations.

The HIL simulation allow the parties involved in the project to test their design and system early in the process and make the necessary modification where needed.

7. REFERENCES

1. J.C. Kalis, D de Koningh, On the Functional Design of an Offshore Wind Maintenance Support Vessel, conference paper RINA Design & Operation of Wind Farm Support Vessels, 2014

2. T.W. DeNucci, 'Capturing Design Improving conceptual ship design through the capture of design rationale', Doctoral Thesis ISBN 97-89065622945, 2012

3. PHILIPP A. WOLFF, 'Conceptual Design of Warships', Doctoral Thesis ISBN 90-36514495, 2000.

4. Marin Report '28493-1-OB - DAMEN – WSV DP MODEL TESTS', 2015

5. Marin Report '27972-1-PO - NUMERICAL DP STUDY Wind farm Service Vessel', 2015

8. AUTHORS BIOGRAPHIES

Jorinus Kalis MSc holds the current position of Manager Development at the R&D department of the Damen Shipyards Group. He is responsible for a team of technical consultants that support Damen's product development, knowledge management and supporting the product groups by providing key technological expertise for the design and production of all Damen products. His previous experience includes several positions within the R&D department of the Damen Shipyards group and a study in Naval Architecture and

Marine Technology with a specialisation in Ship Hydrodynamics.

Sandor Ivancsics is a research engineer at the DAMEN Shipyards Group, specialising in control and automation systems. Studied electrical engineering with a System and Control master track at Budapest University of Technology and Economics. The studies has been finalised in the Netherlands at TU Delft. Sandor is engaged in developing a novel hardware in the loop concept at DAMEN whilst supporting any fundamental or practical issue regarding his field of expertise.

DP POSITIONING – RELATIVE POSITIONING GOES TARGETLESS

A Knight, S Heriot, Guidance Marine, UK
H Busshoff, Bernhard Schulte Ship Management, Germany

SUMMARY

This paper discusses RangeGuard Monopole, the first local DP reference sensor system for offshore windfarms that operates without dedicated targets. Latest results from the first fully integrated system on the SOV Windea La Cour are presented. This paper follows on from a previous paper "A New Era in Local Position Referencing" which was presented at Design & Construction of Wind Farm Support Vessels, 30-31 March 2016, London, UK[1].

NOMENCLATURE

DGPS	Differential Global Positioning System
DP	Dynamic Positioning
PRS	Position Reference Sensor(s)
PSV	Platform Supply Vessel
SOV	Service Operations Vessel

1. INTRODUCTION

Standard microwave position reference sensors[2] used in offshore oil and gas are not optimised for positioning inside a wind farm. Whereas a platform supply vessel in the oil and gas industry may support just 2-3 platforms, a wind farm service & operations vessel supports as many as 150 wind turbines, each with up to three approach angles. Traditional microwave based PRS require the installation of one or more powered targets at each landing point. The installation and maintenance costs of so many targets makes existing microwave based PRS prohibitively expensive for use in wind farms. This leaves operators dependent on laser and DGPS technology for position holding while deploying a gangway. Laser systems[3] are at risk of false reflections, particularly where personnel in hi-vis jackets are working alongside the reflective targets in the close confines of a wind tower. DGPS can suffer from shadowing, reflections and clock jumps. Under these circumstances there remains a need for a third independent local reference sensor with no common failure mode.

Recently SOVs began utilizing also walk-to-work gangways as DP reference system. A downside is that such a PRS only becomes available for a very short duration once the SOV has reached the final docking position, for approach and departure such a system is not available. While gangway-DP in this form also represents a specific windfarm-application only RangeGuard is available for the entire critical stages of a connection to a wind turbine.

RangeGuard Monopole DP offers a microwave based local reference system that does not require the installation of dedicated targets. This reduces costs and enhances safety.

This paper presents the results of the system when compared to a traditional targeted laser system (CyScan).

The trials were conducted on the Bernhard Schulte managed SOV the "Windea La Cour" over the summer of 2016.

Guidance Marine is the only company to offer a local DP PRS solution that removes the requirement for dedicated targets altogether.

2. THE RANGEGUARD MONOPOLE DP SYSTEM

The RangeGuard Monopole system consists of a minimum of two RangeGuard microwave marine range finding sensors mounted onto the side of the vessel, pointing in the direction of the transition piece at the bottom of the wind tower. Simple geometry then determines the range and relative position of the vessel to the transitions piece (as shown in figure 1). The transition piece thus acts as the target to which the system will report range and bearing.

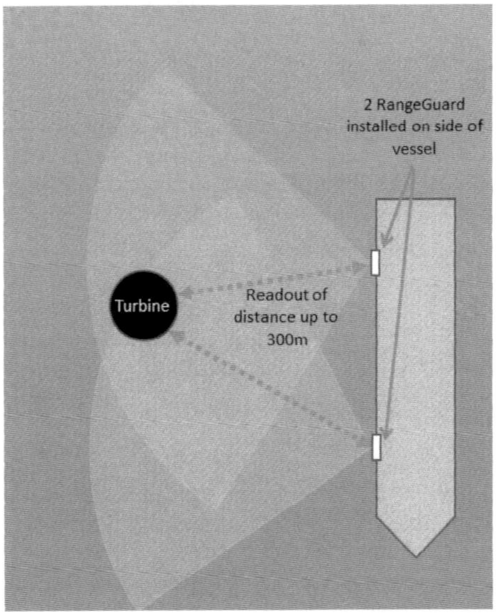

Figure 1: RangeGuard Monopole geometry

The sensors are FMCW radar transceivers operating in the licence free 24GHz band with a maximum working range of 300m[4]. They are configured such that they have an azimuth beam angle of 110° and a vertical beam angle of 11°. Individually each sensor reports the range to the nearest object. Simple wind tower structures offer a single unambiguous object to which both sensors can measure range. Working in tandem they provide the relative position of the vessel which is displayed on the Guidance Marine Dashboard user interface and can be used in the DP system of the vessel.

3. SEA TRIALS

The first sea trials of the RangeGuard Monopole system were undertaken in August 2015 on board the wind farm service vessel Ocean Zephyr[1] managed by Bernhard Schulte Ship Management (Deutschland) GmbH & Co. KG. These successful trials at the Bard 1 offshore wind farm showed that the sensors could accurately report changes to the range and bearing of the vessel relative to the wind turbine where the sensors had clear sight of a simple monopole. Subsequently Bernhard Schulte have installed a full DP enabled system on their SOV Windea La Cour (figure 2) and are in the process of introducing it to the DP PRS weightings.

Figure 2: The Windea La Cour[5]

Sensors were installed on the vessel at the yard on the starboard side of the vessel – the same side as the motion compensated gangway, see figure 3.

Figure 3: Sensor installation on the Windea La Cour.

Guidance Marine accompanied Bernhard Schulte Ship Management (Deutschland) on the Windea La Cour commissioning trials during July 2016, prior to its charter. During a 3-week period, the Windea La Cour visited the Gemini offshore windfarm in the North Sea where the vessel has recently commenced long-term employment.

During these trials, the RangeGuard Monopole system was tested in different scenarios at wind turbines, including approach, departure, station keeping, and gangway deployment. Although the sensors were outputting a DP feed to the Marine Technologies DP system, the DP system was not using the data at this stage. This however allowed the Guidance Marine engineers and the crew to view the data in real time and data was collected for later analysis.

The Dashboard user interface allows the position to be calculated relative to any point on the vessel. For the purposes of comparison with the CyScan the reference point was aligned with the CyScan location across the horizontal. In practice the DPO found it particularly useful to use it with the datum point set at the base of the gangway. Figure 4 shows an image of the Dashboard user interface, which will be familiar to a crew who already use CyScan, RadaScan or Artemis.

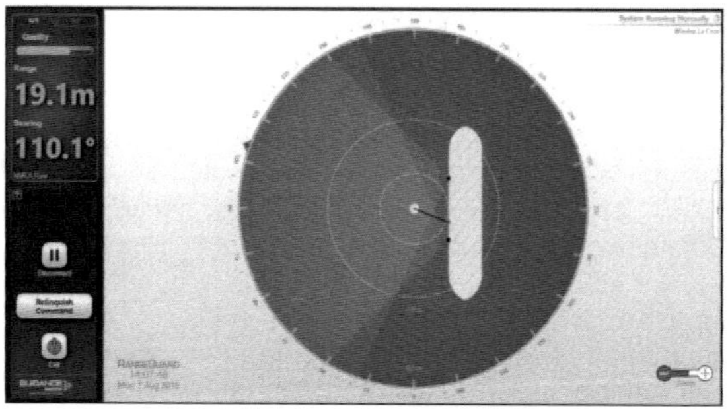

Figure 4: Dashboard user interface. The yellow dot in the centre represents the wind turbine. The black dots show the position of the sensors on the vessel and the red dot shows the position that the range and bearing is measured to. On the left, the range and bearing are displayed.

4. RESULTS

Figure 5 shows the view from the aft senor of a wind turbine transition piece (under construction). It can be seen that the structure is actually more complicated than a simple cylinder. Depending on the angle of approach, the access ladder (to the left of the monopole in this image) may be in view of either one, or both RangeGuard sensors. This introduces the possibility of the sensors first reporting the distance to the ladder but then switching over to the turbine face which would result in jumps in calculated range as the vessel moves.

Figure 5: Aft sensor during sea trial.

After further data analysis, the tracking algorithms have been modified to take into account these features so that a stable range measurement is always reported. It should be noted that for more complicated non-uniform structures, such as an oil platform, this current system could not be used reliably.

During the trials the CyScan laser position reference sensor was also in operation. The CyScan is also mounted on the starboard side of the vessel, figure 6.

Figure 6: CyScan Sensor on the Windea La Cour. The CyScan is slightly higher than the Motion compensated gangway, which is 12-13m higher than the RangeGuard sensors.

Four reflective targets were mounted around the transition piece and position data was collected from the CyScan sensor which selected one of these targets to lock onto. This enabled the CyScan position data to be compared directly to the RangeGuard Monopole position data.

Figure 7 shows the CyScan and RangeGuard Monopole data during a long range approach manoeuvre.

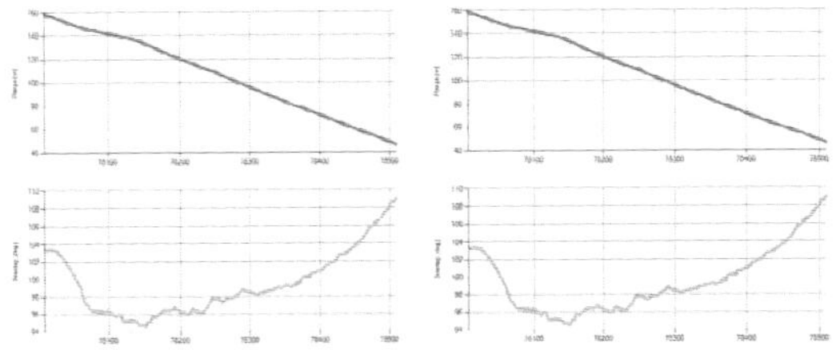

Figure 7: Long range approach manoeuvre. CyScan left, RangeGuard Monopole right, over the same time period. Blue (top) graphs show the range (m) vs time (data counts). Orange (bottom) graphs show the relative bearing (deg) vs time (data counts).

It can be seen that the range curves for the RangeGuard Monopole match very closely to the CyScan data. The bearing curves also match but we can observe that the bearing stability of the RangeGuard Monopole data improves as we move closer in. This is because RangeGuard Monopole uses trilateration to calculate bearing where so noise is greatest where the distance from the target is significantly different to the distance between the sensors. Critically, throughout the approach, the system provides sufficiently accurate positioning data for vessels to meet DP Class out to the 100m DP operation zone.

Figure 8 shows station keeping at the wind turbine, followed by a manoeuvre round to the other side of the turbine (90 degrees from the initial position).

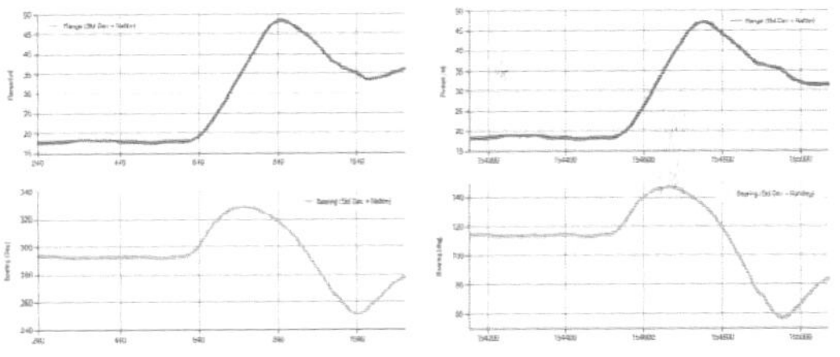

Figure 8: Departure manoeuvre. CyScan on left, RangeGuard Monopole on right for the same time period. Blue (top) graphs show the range (m) vs time (data counts). Orange (bottom) graphs show the relative bearing (deg) vs time (data counts).

Again, there is excellent correlation between the CyScan data and RangeGuard Monopole data. The difference in the range measurement towards the end of the manoeuvre is attributed to the difference in the positions of the targets (CyScan reflector vs. transition piece centre).

Figure 9 shows RangeGuard Monopole and CyScan during station keeping. During this station keeping exercise, the CyScan data and RangeGuard Monopole data match very well.

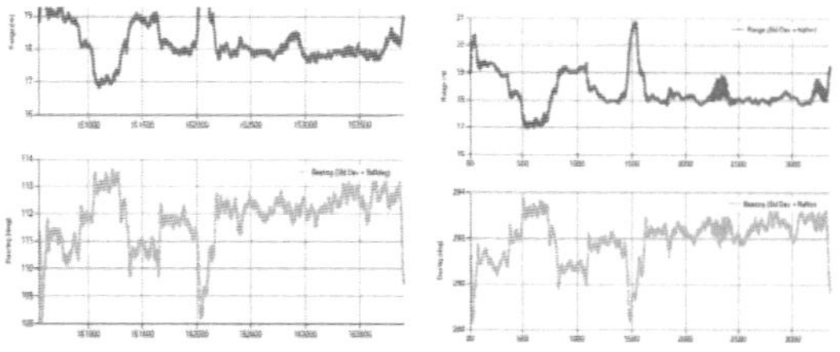

Figure 9: Station keeping. CyScan on left, RangeGuard Monopole on right for the same time period. Blue (top) graphs show the range (m) vs time (data counts). Orange (bottom) graphs show the relative bearing (deg) vs time (data counts).

The x-y position data can also be compared. Figure 10 shows a manoeuvre which includes a period of station keeping.

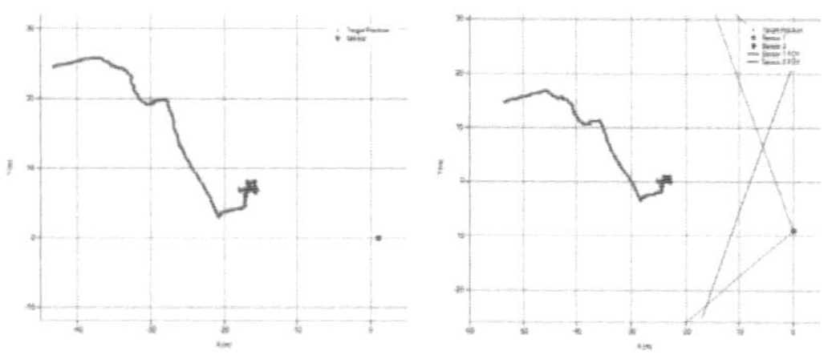

Figure 10: x-y plot (m) showing a manoeuvre which includes period of station keeping. CyScan left, RangeGuard Monopole right. The green line shows target position which correlates to the movement of the vessel in the x-y direction relative to the sensors.

It can be seen that there is little deviation of the relative positions calculated by the RangeGuard Monopole and CyScan. Taking a closer look at the period of station keeping shown figure 11, we see that the CyScan and RangeGuard Monopole traces match very well on the sub meter scale.

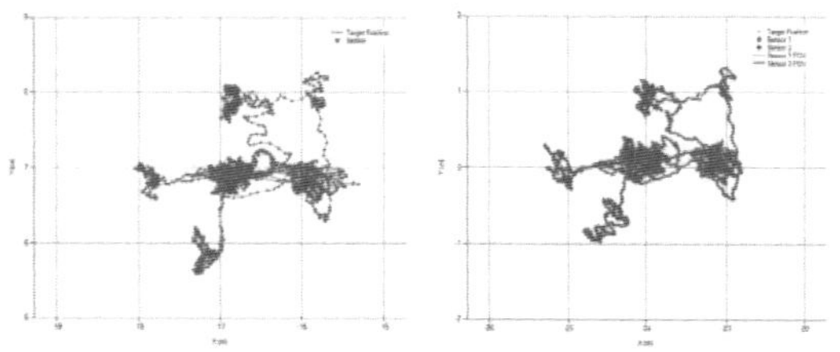

Figure 11: x-y ploy showing the zoomed in period from figure 10 where the vessel was station keeping.

Figure 12 shows a CyScan and RangeGuard Monopole x-y plot data overlaid for another station keeping operation.

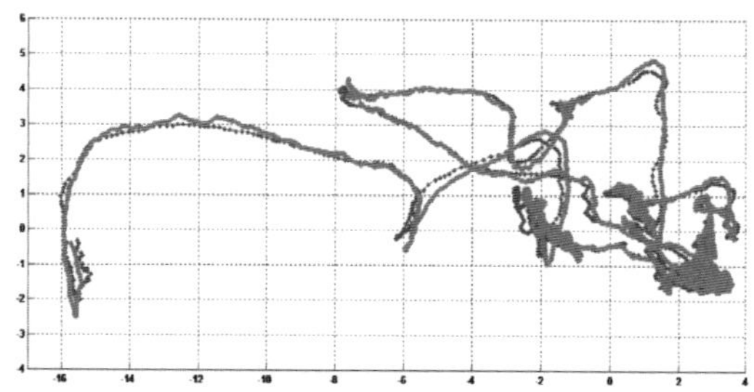

Figure 12: x-y plot (units in m) showing a manoeuvre. The red line represents the RangeGuard Monopole data and the blue line represents the CyScan data.

The slight deviations can be attributed to:

- Vessel motion – there is no motion compensation in these results.
- Errors in RangeGuard Monopole/CyScan measurements – for RangeGuard Monopole this includes range measurement errors caused by the "complex" structure of the transition piece.
- Errors in the model of transition piece – in particular its radius.
- Errors in the mounting position of the RangeGuard sensors relative to system configuration values – for this data set the position of the sensors were estimated from the ship schematics, ideally the position would be derived from survey measurements.

As stated above, the data presented here does not compensate for any roll or pitch of the vessel. These vessel movements can actually be seen in figure 13 which is a zoomed in part of figure 12, squares 0-2 in the x-direction and 1-4 in the y-direction.

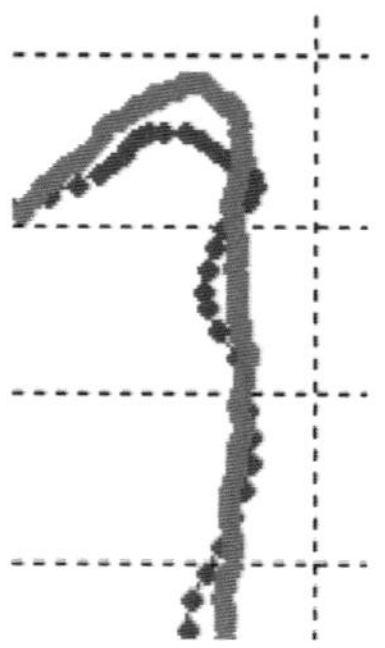

Figure 13: x-y plot (units in m) showing the zoomed in part of figure 12, squares 0-2 in the x direction and 1-4 in the y-direction.

The CyScan data (blue line) in this region is undulating which corresponds to the pitch/roll of the vessel. The RangeGuard Monopole is less affected by this as the sensors are mounted much lower down and closer to the centre of rotation of the vessel. It should be noted that the vessel roll/pitch errors are not only a result of wave motion. For example, they can also be caused by listing of the vessel when the gangway is rotated out over the side of the vessel and when the gangway is pressing up against the transition piece of the wind turbine. This would create a sustained position difference between RangeGuard Monopole and CyScan until the gangway is disengaged/rotated back around rather than the oscillating type error that wave motion causes. These differences are normally corrected in the DP through vessel motion compensation.

5. DISCUSSION

These latest trials demonstrate the viability of using the RangeGuard Monopole PRS against simple unambiguous monopoles. The team at Guidance Marine are working closely with Marine Technologies and Bernhard Schulte Ship Management (Deutschland) to implement the final part of the project and integrate the DP feed into the DP system. This is a carefully defined process whereby small steps will be taken in order to safely realise the capability of the system. Given the results presented here, the team is in no doubt that the RangeGuard Monopole PRS is capable of performing the same function as existing PRS such as CyScan and RadaScan.

It should be noted the DP string provided is a similar format to all other position reference sensors, outputting range and bearing information. It can therefore be easily adapted to any DP system that accepts laser or radar position reference sensors. It is not limited to the Marine Technologies DP system.

Because this is a completely new type of PRS, Guidance Marine is in discussions with the class societies to ensure that the functionality and capability of the sensor system supports vessel DP classification.

Finally, it should be reiterated that at present the RangeGuard Monopole system is only applicable to an environment where the scene contains a single monopole target.

6. CONCLUSIONS

The Windea La Cour is the first vessel to be equipped with RangeGuard Monopole, a new type of PRS designed specifically for offshore wind DP operations. Based on radar technology it is unique in the fact that it does not require any dedicated targets located on the wind turbine. Instead it uses radar reflections from the environment to measure range to the nearest objects in its field of view. As an additional PRS sensor which complements traditional PRS, this has potential to significantly reduce cost and improve safety during offshore wind operations, as well as providing an additional form of redundancy to the DP system. The common problems that occur with mounted targets are negated and maintenance of targets is no longer required.

Due to the success of the results and collaboration between Guidance Marine and Bernhard Schulte the sister vessel to the Windea La Cour, the Windea Leibnitz has now had RangeGuard Monopole installed. The system is expected to have Type Approval granted by April 2017.

7. ACKNOWLEDGEMENTS

Thanks go to the crew of the Windea La Cour who have and continue to cooperate throughout the trials, the Ulstein Shipyard who installed the sensors on the vessel, the team at Bernhard Schulte Ship Management (Deutschland) who welcomed us onto the vessel during commissioning and the team at Marine Technologies who are working on implementation of the DP feed into the DP system. Finally thanks go to the rest of the team at Guidance Marine who have been involved in this project to make the first targetless sensor!

8. REFERENCES

1. SY Heriot, T Coggins, M Mudrinic & A Stead, H Busshoff, O M Husøy, A New Era in Local Position Referencing, Design & Construction of Wind Farm Support Vessels, 30-31 March 2016, London, UK

2. Guidance Marine, http://www.guidance.eu.com/radascan

3. Guidance Marine, http://www.guidance.eu.com/cyscan

4. Guidance Marine, RangeGuard Proximity Sensor, *http://www.guidance.eu.com/assets/_managed/cms/files/New%20Brochures/94-0430-4-A%20RangeGuard%20Brochure..pdf*

5. *David Foxwell, Offshore Wind Journal, Position reference sensor is ideal for offshore wind application, 10 Aug 2016 http://www.owjonline.com/news/view,position-reference-sensor-is-ideal-for-offshore-wind-application_44087.htm*

9. AUTHORS BIOGRAPHY

Andy Knight is a Senior Software Engineer at Guidance Marine and has been the lead Engineer on the RangeGuard Monopole project. He has been with Guidance Ltd since 2007 and his previous experience includes working on RadaScan PRS software, radar/laser signal processing and position estimation for automated guided vehicles.

Sasha Heriot is Business Development Manager at Guidance Marine Ltd. After completing her PhD in Physics at the University of Sheffield in 2003, Sasha worked in the field of polymer physics as a research associate for 4 years. She left the world of academia to work in a spin out company from the University of Leeds specializing in novel surface treatments. She then worked for 3 years in a multinational chemical company as Market Development Manager for a specialist chemical business unit. Sasha started her appointment at Guidance Marine in April 2015 where she is currently introducing innovative technology into new markets and new applications.

Hendrik Busshoff holds the current position of Offshore Marine Superintendent at Bernhard Schulte Ship Management. As a trained Master mariner he is responsible for offshore marine operations and DP. His previous experience includes SDPO on hyperbaric dive support vessel in the North Sea with Technip and sea time as officer and Marine Superintendent with the merchant navy.

AUTHORS' CONTACT DETAILS

John Haynes
Shock Mitigation Ltd
john@shockmitigation.com

Jorinus Kalis
Damen Shipyards
jorinus.kalis@damen.com

Mike Longman
Nauti-Craft Pty Ltd
alex@nauti-craft.com

Russel Edson
Withers & Rogers LLP
redson@withersrogers.com

Sasha Heriot
Guidance Marine
sasha.heriot@guidance.eu.com

John Kecsmar
Ad Hoc Marine Designs Ltd.
john_lol@btinternet.com

Matthew Shanley
University College Cork
matthew.shanley@ucc.ie

Oleksandr Bondarenko
Admiral Makarov National University of Shipbuilding
abondarenko70@mail.ru

Stephen Phillips
Seaspeed Marine Consulting Ltd
stephen@seaspeed.co.uk

Silvie Votrubova
PBES
svotrubova@pbes.com